Finding the Light

Finding the Light

SCIENCE AND ITS VISION

DR. JAMES A. CANNON

ALPHONSUS

Xulon Press
2301 Lucien Way #415
Maitland, FL 32751
407.339.4217
www.xulonpress.com

Exulon ELITE

© 2023 by Dr. James A. Cannon

All rights reserved solely by the author. The author guarantees all contents are original and do not infringe upon the legal rights of any other person or work. No part of this book may be reproduced in any form without the permission of the author.

Due to the changing nature of the Internet, if there are any web addresses, links, or URLs included in this manuscript, these may have been altered and may no longer be accessible. The views and opinions shared in this book belong solely to the author and do not necessarily reflect those of the publisher. The publisher therefore disclaims responsibility for the views or opinions expressed within the work.

Unless otherwise indicated, Scripture quotations taken from the New American Bible Revised Edition (NABRE). Copyright © 2010, 1991, 1986, 1970 Confraternity of Christian Doctrine, Inc., Washington, DC All Rights Reserved.

Paperback ISBN-13: 978-1-66286-578-7
Dust Jacket ISBN-13: 978-1-66286-579-4
Ebook ISBN-13: 978-1-66286-580-0

Table of Contents

Chapter 1: Personal Faith 1
  The Human Spirit 1
    The Religious Sense. 2
    The Religious Experience 4
  Blaise Pascal (June 19, 1623–August 19, 1662) 8
    Normandy Years, 1640–1650 11
      The First Conversion. 13
    Back to Paris. 15
      The Second Conversion 17
  Alexis Carrel (June 28, 1873–November 5, 1944) 23
    Vascular Surgery 24
    Rejection and Fulfillment. 26
    Angst at Lourdes. 26
    Early Religious Ambiguity 30
    Carrel's Unique Position. 31

Chapter 2: The Emergence of Modern Europe 35
  The Beginnings of Science. 35
    The Feudal System. 35
    Medieval Scholarship. 37
      The Beginnings of Scientific Thinking. 37
    The Role of Philosophy 39
      What Is Philosophy? 40
      The Development of Scholasticism 41
      Philosophy as an Aid to Understanding. 42

  The New Philosophy of René Descartes 43
    The Philosophical Logjam 44
    Reductionism 45
    Deep in Idealism 46
    The Idealist Trap. 49
  The Scientific Revolution. 50
    The Emergence of Classical Physics 50
    Isaac Newton's Physics and Mathematics 54
    Pascal's France 56
    Between Two Lives. 58
  Appendix: Newton's Three Laws of Motion. 59

Chapter 3: The Loss of Christian Unity 61
  The Renaissance. 61
    The Spirit of the Renaissance 63
  The Reformation 65
    Religious Wars. 65
      The Saint Bartholomew Massacre 66
  Mère Angélique and Port Royal 67
    Jansenism 70
      The Suppression of Jansenism 72
      The Provincial Letters 75
  The French Revolution 76
    The State of the Nation 76
    The Course of the Revolution, 1789–1795 78
      The National Assembly. 78
      The Assault on the Church 79
      The Civil Constitution of the Clergy (July 12, 1790). 82
      Regicide and the Reign of Terror. 84
      Empires, Monarchies, and Republics, 1795–1871. 85
      Science in Carrel's Time. 86
  P.S. The Light of the World 87

Chapter 4: A Brief History of Physics ... 89
Nineteenth-Century (Classical) Physics ... 89
Advances in Mechanics ... 89
Michael Faraday ... 90
Basics of Electricity ... 91
Magnetism ... 93
Electric Force ... 94
Faraday's Models ... 95
Electric Flux Density ... 95
Magnetic Force ... 96
Electromagnetic Induction ... 97
Maxwell's Electromagnetic Waves ... 99
Thermodynamics and Statistical Mechanics ... 101
Twentieth-Century (Modern) Physics ... 103
Max Planck and the Black Body Problem ... 103
Albert Einstein's Miracle Year ... 104
Planck's Quantum and the Photoelectric Effect ... 105
Space-Time and Relativity ... 106
Appendix: Time Dilation and Length Contraction ... 109

Chapter 5: Science and Other Ways of Knowing ... 117
Pascal's Epistemology ... 117
Pascal's Three Orders ... 119
Describing the Orders ... 121
What Is a Proof? ... 124
Carrel at His Peak ... 127
Carrel's Religious Sense ... 127
The Rending within France ... 128
Epistemological Restrictions ... 129
Christian Scientism ... 130
The Study of Miracles ... 132
Collaboration with Lindbergh ... 134

Rejection of Descartes 136
    Non-specialist Scientists. 136
    Obstacles 137
P.S. Doctor and Patient 137

Chapter 6: Thinking as a Realist 139
Current Science and Philosophy 139
    Modifications from Einstein's Relativity. 139
    Quantum Physics and Its Modifications. 140
        The Principle of Indeterminacy 142
        Other Quantum Experiments 143
    Carrel in Retrospect. 144
        The Unique Strengths of Youth 144
        The Loneliness of Being Unique. 145
A Case for Metaphysics 148
    Philosophic Roots of Science 148
    The Ontological Structure of Being 149
        Transcendent Characteristics of Being. 150
    Freedom and Culture. 156
        The Person and Practical Reason 156
        The Culture as Context 159
        Thomistic Realism 160

Afterword 163

Introduction

The light that has been lost is faith. Some indications of this are the popular books written by scientists in recent years debunking religious faith in the name of science. With a disdainful air, some take faith as an affront. These atheists are aghast that anyone in this day and age could be so naive as to believe in God. How should the layman react to this? What is going on? How did we get to this point?

It did not happen out of the blue. In the fourth century B.C., Aristotle authored a philosophical treatise, *On the Heavens*, which was the primary classical reference on the universe from his day until the early fourteenth century. Aristotle's description of the universe contained no sense of experiment.

The birth of science stalled for eighteen centuries until John Buridan, holder of a chair of philosophy at the Sorbonne, commented on Aristotle's *On the Heavens*. When he read *On the Heavens* in the early fourteenth century, he approached it critically, examining Aristotle's ideas in light of what his own experiments on motion had shown. He had deduced that space, time, matter, and motion were universal realities. He looked for evidence of experiment and found none. He had no reason to defer to Aristotle. He can reasonably be credited with endowing experiment as central to physical understanding.

Prior to the Christian culture of Buridan, other cultures, for various reasons— superstition, religion, politics, philosophy—were unable to initiate science until Buridan's culture provided the conditions necessary.[1]

---

[1] See Stanley L. Jaki, *Science and Creation*, (Insert publication city: Scottish Academic Press, 1986), 232–35.

Buridan lived when Europe was emerging from its medieval past. The Renaissance roughly extended from the early fourteenth century in Italy to the early seventeenth century. It was a time of cultural, economic, and political expansion. Cities, most notably Venice, became international trading centers. Tradesmen of many types set up businesses and prospered, providing money for investment in the arts and architecture. The Renaissance was going to change European culture drastically.

Remnants of feudal culture persisted in the social classes, the church, the nobility, and the commoners. Some nobles had titles bearing benefices, income, without doing any work, like a government no-show job. In addition, in France the monarchy was constantly expanding its power, spending profligately and raising taxes. People meanwhile coped with the imperatives of their day, living in the real world.

While political and economic changes were more evident (e.g., a growing wealthy class), philosophic changes (e.g., Cartesian thinking) were potentially more infectious.

Two developments became significant: (1) the emergence of experimental science as a model of intellectual achievement, and (2) the rejection of medieval philosophy due to its identification with churchmen in the Galileo (1564–1642) affair.

Science was perceived as an antidote to the churchmen and their philosophy. The censure of Galileo by the church precipitated an apparent antipathy between science and religion, even though virtually all of the Renaissance scientists were Christian.

When the unity of Christianity was broken by the Protestant Reformation, interreligious biases and violence ensued. Also, political upheavals, led by the French Revolution, exposed all of the excesses of French culture. The unfortunate presence of an inept king, Louis XVI, who provided no leadership or patriotic fervor, encouraged the strongest and most organized factions, making it impossible for France ever to be the same. As

Introduction

Europe emerged, scientific challenges to religion became more pronounced with the development of materialistic philosophy.

Returning to an earlier question: How can a layman respond to atheistic scientists? Scientists alone can only depend upon their senses and their minds. Lacking faith, these scientists can only say: "I can't see Him." Such scientists can say nothing about God. For those with faith, no proof is needed. For those without faith, no proof is possible.

Now we will examine the events we have discussed. Think of this as a tour. On a tour many things are meant to be enjoyed. You can pick and choose. There is a story being told. There are many interesting people, and then there is science to encounter as you may.

You will be guided historically, covering science from Buridan's beginning to Einstein and beyond. You will also encounter much on the importance of philosophy, the love of wisdom. You will meet diverse and interesting characters whose biographies are varied and unique. Absorb as much science as you can and then move on, or go into science as deeply as you wish. You can go back to details later. There is plenty of interest in the biographies, and the history is compelling. Enjoy the tour. You will encounter many very interesting people and ideas.

Chapter 1

Personal Faith

The Human Spirit

Human spirits are unique in that we are conscious of ourselves and, through language, conscious of other human spirits. We are spirits engaged in finding light, knowing! Science is one way of knowing, but its "vision," its ability to see, has limits. We can study how things present themselves to us and how they relate to each other. We discover principles by which we understand things. We can imagine possible things. In normal circumstances, we are capable of knowing when there is a reality before us.

Science is a set of disciplines devised to deepen human understanding of the physical world. Its methodology depends upon experiment—specific sense knowledge of the object studied.

The need to survive drove our forebears to seek first their personal security and that of their community. Apprehending the workings of the environment is the first imperative in enabling man to provide for his needs and those of his community. A community's decisions, over time, set customs that influence the growth of culture and the formation of law. Common development begets a character and a history that are passed on to future generations.

Human beings are capable of acting freely and of observing themselves acting. We are aware of ourselves interiorly and we observe what particular understanding prompts our action or inaction. This human self-awareness infers a responsibility for man's motives and actions.

Any mature community, culture, or civilization will evolve by virtue of the use of these two intangible, nonphysical human capacities: the ability to know and the ability to choose. These are spiritual capacities and imply the importance of truth and goodness.

Finally, we know that any given person may make choices incompatible with the good of the community. The possibility, indeed the certainty, of unacceptable choices requires every community to formulate its own ethic.

The Religious Sense

Inevitably, by virtue of its particular development, each society has some religious tradition. Each society recognizes that at the root of being, there are mysteries, awe-inspiring unknowns. Manifestations of these understandings are usually seen in the rituals associated with basic human passages—birth, puberty or coming of age, family customs and, ultimately, death. Such common human experiences provoke wonder—something beyond themselves.

Sociologists can use religious rituals as a window into beliefs and values. Each human being, as a member of the community, is free to internalize such beliefs and values to whatever degree he wishes. They are in the air he breathes. A disposition to recognize this dimension of human experience is what we will refer to as a religious sense. A religious sense does not necessarily imply that a person believes in God. But it does mean that a person recognizes that he is subject to wonder in the face of his existence.

This realm of wonder clearly is a soft light often unexamined in the press of imposing cares. In some, it rises rarely and may fail to compel any response. Many of us do as we please in the ordinary course of our days. But when tragedy strikes—either to a person or to the community—the religious sense awakens. For example, in the hours immediately after the 9/11 attack, in the nation at large but especially in New York City, people flocked to churches. The first impromptu services in the city that day were standing room only. Even politicians could be seen standing with the masses. The news for many weeks was dominated by formal national services and local services, mainly religious funerals.

Many who respond to the religious sense come to a conviction regarding God and follow a personal faith. Religious faith is a radically different way of knowing—in the sense that its radices, its roots, are different. Many throughout history have written testimonies of the experience of God. Religious experience, as I will use the term, is one of these individual insights, as opposed to the religious sense, which is a common inclination.

The conflict between science and religion arises in the mind of some so-called materialists because they claim that there is no basis "scientifically" for religious ideas. This is true. However, it is true because, as a matter of method, science can only be applied to situations where experiments can be performed. Sense information supplies the evidence and the vindication of science. Therefore, to examine a phenomenon that is not physical, such as religious experience, is beyond the domain of science. The manner of God's presence to the believer does not lend itself to measurement.

On the other hand, reason demands that religious people respect the validity of properly conducted science. A scientific theory supported by consistent experimental evidence and subject to falsification by new experiments should be respected as the best current understanding. There are always scientists who do not agree with the prevailing wisdom, who think "outside the box." Einstein is a good example. Some scientists question certain aspects of modern physics. But their dissatisfaction is based on well-articulated reasons. While they accept the experimental data, they may question its interpretation. Such caution prevents overreaching and is the basis for continued testing. The scientific community will replace or augment a theory of long standing when new experimental evidence demands it. Generally, such action leads to a greater unification of understanding.

Modern materialists, as a matter of method, reject any evidence that is not experimental. The rejection is ideological. They expect a believer to deny his own experience because it is outside their self-designated realm of truth. This doctrine, that science is the sole means to truth, is called scientism. We will discuss the nature of these problems in a later chapter.

Despite its persistent lurking in human sensibility, the religious sense is denied by some. A person who cannot decide one way or the other about God is an agnostic. A person who outright rejects belief in God is an atheist. G. K. Chesterton has observed that when one doesn't believe in God, one will believe in anything. If there is no God, the latent religious sense is impelled to seek other objects of worship. When a state is founded on secular or atheistic ideologies, these pale substitutes for God become the oppressive imperatives of totalitarian politics. There is no basis for a humane jurisprudence or for democratic tolerance in dictatorial governments. Nazism and Communism ultimately failed in the twentieth century because they could not crush the religious sense in man.

THE RELIGIOUS EXPERIENCE

Religious experience describes the multifaceted ways God reveals Himself to individuals. To begin, how does God manifest himself to a believer? Prayer is how man addresses God. It is common to all faiths and even has been tried by atheists in foxholes. This believer believes there is one God, a personal God, who hears the prayers of all. I believe that God is active in every human life. He maintains each of us in existence and loves us individually. Creation is a gift from Him. He creates us with a free will. We may reject Him or ignore Him. He may hound us, but He will never compel us. Maybe the animus of atheists toward God is due to their inability to shake Him. On the other hand, He will respond lovingly to anyone at any time, at any level of interest, even the smallest opening: As the Psalmist says, "a contrite, humbled heart, O God, you will not scorn."[2]

The God of Western history, inferred by the Greek philosophers as the God of the Jews and Christians, is a personal God, the creator and maintainer of everything that exists. Judaism, Christianity, and Islam profess a belief in one God, who is a person. Yahweh, the God of Abraham, Isaac, and Jacob, revealed himself to Israel through Moses and the prophets. He selected the Jewish people to be His own. In the Torah, the Law, we have the basis for Yahweh directing the Jews throughout their history. Christians believe that Jesus Christ is the Son of God, who came into the world in fulfillment of the promises made to the Jewish people. He wishes to bring all men, Jews and non-Jews, to eternal life.

There are philosophical arguments for the existence of God, and we should be aware of them. Such approaches to God are framed indirectly by seeking an explanation for the totality of reality (including the moral sense) that is evident to man. The God of the philosopher is evident, is implicit—hidden but reasonable.

So how does the believer know God? What does God say, and how does He say it? Since God is a person, how does one get to know God personally? Well, God does not appeal to our intellects, primarily. He can be discursive with us, but usually He is hidden. He short-circuits our intellect, not seeking to prove anything. But He does reveal Himself in a radically direct way. He loves us.

He appeals to our hearts. He is immediately present to us. He invites us to move day-by-day to a closer unity of our will to His will. He wants us to grow day-by-day to greater knowledge of Him. Experiencing God's love is not a logical demonstration or theorem. By loving our neighbor as ourselves, we grow in His love. There is no end to this love. There is an enduring spiritual exclamation point, a level of assurance and confidence, that exists

---

[2] Ps. 51

amidst misery and good fortune. He is open to everyone but especially the poor and the humble. "He has thrown down the rulers from their thrones but lifted up the lowly."[1]

The believer knows God by His presence in his life. As a child, he may learn to pray and learn about God's love via parental faith and religious practice. But at some point the youngster must personally assent to or reject the faith of his parents. When a person assents, it is God's presence that he avows. God loves first. Faith, living in His love, is a gift of God to the person.

We may reject the gift, but if assent occurs later in life, under different circumstances, it is because God has been prompting us. We must respond by loving Him freely. His appeal is to our will. To return His love is a decision that cannot be compelled. He wants us to unite our will to His will. His will for us is eternal life. Doing the will of God for each believer is a lifelong task.

The route is never static. There may be times of elation (consolation) when God is close and times of loneliness (desolation) when God appears to be absent. Such trials are opportunities for growth in His love. When media reports on the later life of Mother Teresa revealed that she did not experience consolation for many years, the news was misinterpreted as a spiritual flaw. Desolation is a spiritual trial along the road to sainthood. She clearly did not stop loving Him. She continued to do His will as she understood it.

What we have been referring to as His presence, His loving, and His drawing us to Him is called *grace*. *Grace* is a name for the infinitely varied gifts of love that He offers. The gifts given are all custom-made. One size fits one. God loves each of us in a unique way, personally. The experience of this love is direct—freely given. There are no syllogisms. You have nothing to analyze. Indeed, intellectualizing gets in the way. God's manner of working is unique to the person. The experience is transcendent, beyond comprehension. It is a communication between two free spirits—a dance, a courtship, a divine drama.

The late Hans Urs von Balthasar, a noted theologian, was reluctant to offer any intellectual framework to describe this dance. Dance, at its root, is a response to music. Balthasar had a deep appreciation of music, and I believe he saw the human response to beauty in music as a metaphor for the experience of God's presence to the individual soul. Good music is appreciated at the hearing; no explanation is needed. The awareness of God to the prayerful soul is spontaneous. The soul seeks unity with God.

An extensive mystical literature has provided commentaries throughout the centuries on how God works in the individual soul. *The Cloud of Unknowing*, by an anonymous fourteenth-century English monk, teaches that God's presence may be found by striving to remove all mental and physical distractions, thus creating a quiet openness for the soul within which God may be heard. The cloud of unknowing is a recollected state, devoid of thought, within which the promptings to the soul may be discerned.

Of similar nature are the writings of the sixteenth century Spanish Carmelites, St. John of the Cross and St. Teresa of Avila. The writings of St. John of the Cross are poetically exquisite. *St. Augustine's Confessions* is a classic of both literature and religious conversion.

Another marvelous account of personal experience with God is *A Revelation of Love,* by a fourteenth-century English mystic, Dame Julian of Norwich. She chronicles sixteen showings, of revelations and conversations with Christ on his passion. One very prominent characteristic of Christ in the showings is His abiding courtesy.

There are relatively modern books such as *The Story of a Soul* by St. Thérèse of Lisieux, published in 1899 and never out of print. St. Thérèse was allowed to enter the Carmelite convent at fifteen. In frail health, she died nine years later. Thérèse's account of her struggles with loving as God loves and with depression is a marvel of simplicity and power.

The awareness of God's presence is seldom experienced in special incidents, but the believer comes to recognize his whole life as a continuum of experience within which God appears to be moving us, day by day, as an abiding light. Dramatic experiences are rare. However, St. Paul certainly got special attention on the road to Damascus, when he was knocked to the ground and scolded.

Some might experience "out-of-the-blue" intuitions. Overwhelming joy may arise spontaneously as part of the divine dance, but most often one is aware of God's presence in the living of one's life. This ordinary awareness may be heightened—one's heart may be lifted up by the reading of scripture or by prayer, by worship, or by an encounter with another person.

The faithful person aware of God's love responds, but the response may be tempered by the noise of the world. As a person lives on, he should see more clearly how God has worked in his life—how he has been led and prompted. God is present to the person in the routine events of life, in the challenges encountered, in joys and sorrows. Most who seek to do His will experience a bumpy but steady growth in awareness of His presence and can reflect on how he has been steering them. This life of deepening commitment may have marked times of consolation and desolation consistent with the whole of life.

In the absence of faith, God will not interfere with one's choices. If the promptings of the heart are ignored, if the noise of the world prevails, God will remain hidden. At every point God respects our free will. But He is an untiring lover. He will offer his love to the end. The best-known late bloomer is the good thief, who won salvation on the cross.

Let us now look at two scientists who have recorded their religious experience explicitly. They differ significantly. They are separated in time by more than 210 years, during which the cultural environment changed drastically.

BLAISE PASCAL (JUNE 19, 1623–AUGUST 19, 1662)

Seventeenth-century France was one ground within which the economic, political, and philosophic seeds of modern Europe germinated. Internal political unrest and the cost of incessant wars accompanied the growth of nation-states. In addition, the Protestant Reformation had catastrophic consequences for European culture.

However, international economic and social interaction flourished, particularly with the Moslem and Byzantine worlds. Craftsmen filled cities as new economic opportunities arose. A class of wealthy commoners, the *noblesse de robe,* augmented the landed aristocracy, the *noblesse d'epée.* These wealthy people who purchased titles of nobility and privilege were a source of income and provided competent infrastructure for the monarchy. Many used their money and time for education and patronage of the arts.

On June 19, 1623, Blaise Pascal was born to Etienne and Antoinette Pascal (née Begon) in Clermont-Ferrand in the province of Auvergne, about two hundred miles south of Paris. Etienne's father, Martin, had purchased a position in government service for which his experience had prepared him. His *noblesse de robe* status brought income and influence.

Etienne had married Antoinette in 1616,[23] and they had three children. Blaise's elder sister, Gilberte, was born in 1620. She would marry Florin Périer in 1641. Blaise's little sister, Jacqueline, born in 1625, was very close to her brother. She had an exceptionally sweet temperament and a talent for poetry. She would take religious vows in 1651. The family spent the early years in Clermont, where Etienne was a tax court officer for Auvergne. Not much is known about Blaise's mother, Antoinette. When Antoinette died in 1626, Etienne took direct personal charge of rearing and educating their children. He did not remarry.

When Blaise was eight, Etienne moved his family to Paris, where the best tutors were available. Indeed, Blaise never attended an established school. His father supervised his curriculum and raised all his children in a close, nurturing environment.

Etienne, a skilled mathematician, was part of a circle of practicing scientists and mathematicians in Paris. As a teenager, Blaise was able to share

---
[3] O'Connell, Marvin R., *Blaise Pascal: Reasons of the Heart,* W. B. Eerdmans, Grand Rapids, 1997. The primary reference for the facts of Pascal's life is O'Connell's book.

this company and in very short order became an active and valued member of the intellectual class in Paris. Jacqueline's pleasant nature and her skill in verse made her a favorite of the bourgeois and noble families, including the king and queen.

The Pascal family was formally Catholic, but they were not directing their lives especially toward growth in holiness. Etienne was a public man of a practical and intellectual disposition. He did not reflect much on the demands of faith in his personal life. He was, according to the term used, an *honnête homme*, a man of affairs who shared the worldly wisdom and tastes of his class.

Over France's long history, the monarchy and the Catholic Church had become operationally intertwined. The feudalism of the Middle Ages was primarily a defensive network of agrarian communities in which the laboring agrarian class traded their work for the protection afforded by the local land-owning nobility. The Church was a trusted intermediary that would provide social justice in the community.

However, French statecraft, while formally recognizing the faith, put the interest of the state first. As history played out, the state was wary of what they called *ultramontaine* (beyond the mountain) papal influence. This disposition, referred to as Gallicanism, exerted a strong secular influence on the Church, which undermined its pastoral role.

In 1638 Etienne made a major blunder. He complained a little too loudly about the crown's lowering the interest on some government securities. He got word that the first minister, Cardinal Richelieu, was preparing to arrest him for his lack of team spirit. He quickly fled to Clermont-Ferrand, leaving his children in the care of trusted overseers.

Soon thereafter, with the contrivance of Etienne's friends, Jacqueline had a chance to display her special talents with verse and her appealing nature to the court. She charmed all including the king, Louis XIII, and his queen, Anne of Austria. Then, early in 1639, after Jacqueline had recovered from a disfiguring case of smallpox, she participated in a *fête* in honor of Richelieu. Jacqueline charmed all again with her recitation and vibrant presence. She led Richelieu to relent in his treatment of Etienne and asked him to give her father a chance to offer his respect personally. Within a year, Etienne was appointed tax commissioner in Normandy.

Certainly, Etienne's *faux pas* was understandable in personal terms. An arbitrary change in the terms of a contract certainly was unjust, but any challenge to the prerogatives of the monarchy was treated very seriously. The power of the king was a paramount priority for Richelieu. No doubt Jacqueline's talent and personal earnestness softened Richelieu, but he could afford to be magnanimous. Etienne had learned his lesson. He had a wonderful family, and he still could be a valuable servant.

Normandy Years, 1640–1650

In the summer of 1639, civil unrest made it impossible for Etienne to begin his work as a tax assessor in Normandy. The overtaxed citizenry—nobles, merchants, and peasants—were in open rebellion against the crown. It took the better part of a year to quell the resistance. Early in 1640, the Pascals settled in Rouen on the Seine about one hundred miles northwest of Paris. Etienne had been given no plum.

Blaise was sixteen and had just finished an exceptional, though not original, mathematical treatise on conic sections. Professional coolness toward Etienne may have led René Descartes, the renowned mathematician and philosopher, to offer dismissive comments on the work. Yet the work brought attention to the remarkable talent of this youngster. The family was growing up. Gilberte was going to marry Florin Périer next year, and of course, Jacqueline was a very talented and personable teenager.

In Rouen, Blaise became aware that his father had to perform routine arithmetic operations endlessly in order to manage tax records. He set out to create an "arithmetic machine" that could mechanically perform these operations. In 1643, at the age of nineteen, he had made a working prototype.

In 1645 he was satisfied that he had optimized the device, an embryonic computer. He made many different versions. It worked, but he had trouble making it a commercial success. Manufacturing, servicing, and marketing problems were insolvable at the time.

In the summer of 1647, Blaise, who was often ill, was suffering severe headaches. He moved to Paris and focused on studying the physics of a vacuum. Indeed, at the time the possibility of creating a vacuum was questioned on philosophical grounds and had not been tried scientifically. Pascal set out to study the possibility of creating a vacuum—how to produce it, how to demonstrate it was real, and what of its behavior could be measured.

In the fall he completed a paper describing a variety of techniques he had used to create a vacuum. (The following discussion describes one technique and requires some concentration. If you wish, you need not assimilate the details.) One can start by pouring a liquid, most commonly mercury, into a glass tube with a closed bottom. The mercury rests at the bottom of the tube. Next connect a flexible (rubber) tube to the opened end of the glass tube. Then turn the rubber tube to form an inverted U-shaped setup (like an arched doorway) with the mercury at the bottom of the glass at one side. The other side (the rubber end) is exposed to the pressure of the atmosphere. Now invert the entire setup, putting the mercury at the top of the glass tube while the opened (rubber) end is still exposed to the atmosphere. The mercury in the inverted glass tube now falls due to its weight, leaving nothing (a vacuum) above the mercury.

In his *New Experiments Concerning the Vacuum*, Pascal describes eight different techniques by which an empty section is created within a confined liquid. Within the paper he anticipated many objections—answering them in detail. His correspondence with doubters, published with the paper, demonstrated his reasoning. Pascal unrelentingly appealed to the experimental data in justifying his conclusions.

This paper was quickly followed by a second paper, *The Great Experiment on the Vacuum*. The experiment was conducted, following Pascal's directions, by his brother-in-law, Florin Périer, on November 19, 1647, on a mountain, Puy de Döme, in Auvergne. The experiment studied the effect of altitude on a vacuum created in the manner described in the first paper. The device used a U-shaped tube closed at one end, what we would call a manometer today. The experiment studied the vacuum at the closed end as the open end was exposed to the atmosphere at different altitudes on the mountain. At higher altitudes the vacuum at the closed end expanded. The mercury dropped.

Pascal argued that the vacuum expanded because the weight of the air pushing down on the tube at the open end was reduced as the altitude increased. The fixed weight of the mercury was able to expand the vacuum because the atmospheric pressure was reduced. The experiment shows that the pressure of the atmosphere—i.e., the force per unit area of the ocean of air we live in—decreases as the altitude increases. Thus, the vacuum expands as the apparatus goes to higher elevations[4].

In a 1654 paper published after his death, Pascal outlined the implications of his work, especially that a vacuum can be created. There really can be a space with nothing in it.[5] The weight of the atmosphere is the force that pushes air into any empty space, making it so that nature "abhors a vacuum." Pascal's eminence as a pioneer in the physics of fluids is evidenced by the metric unit of pressure being called the *Pascal*.

The First Conversion

Early in 1646 Etienne Pascal slipped on ice and dislocated his thigh. He was forced to convalesce at home for three months. During that time he availed himself of the services of two brothers, the Deschamps, from nearby Rouville. These men had been known as notorious ruffians but exchanged their ways for Christian piety under the influence of their pastor. The

---

[4] If you pay attention to weather reports, the pressure of the atmosphere at sea level is measured as approximately 30 inches of mercury, Hg, or 76 cm of Hg.

[5] Today we would qualify this interpretation to recognize the existence of mercury vapor in the 'empty' space.

brothers challenged the Pascals' *honnête homme* Catholicism—to adopt a greater level of piety under the influence of the pastor, Curé Jean Guillebert.

Three years earlier, this priest-theologian had abandoned academic life to be the pastor at Rouville. He had been a theologian at the Sorbonne, where he had a benefice as pastor of Rouville with a stipend. Benefices were titles conferring an income but requiring no pastoral work, an obvious corruption. When Guillebert, however, decided to attend to his pastoral title, he proved to be a true shepherd, leading a community of devout believers—among them, the brothers Deschamps. These gentlemen lived in the Pascal home for three months. Though they had no professional training, they were skilled in patient care and served Etienne well. But they also let it be known that they were concerned with Etienne's soul as well as his leg.

The brothers challenged the worldly focus of the family. They had a particular interest in young Blaise. Their pamphlets and conversation reflected Curé Guillebert's spirituality, which called for serious devotion to Christian ideals. Blaise was attracted by their enthusiasm. He convinced Jacqueline, who was only marginally interested, to consider their testimony. Etienne admired their charitable work, but his worldview did not allow for such advocacy. However, when Gilberte and Florin Périer responded favorably, Etienne went along.

This family decision is referred to as Blaise Pascal's first conversion. It was a rejection of his *honnête homme* past in favor of a serious commitment to the counsels of his faith. At this time, the affection between him and Jacqueline deepened. She grew in commitment to the devout life. Blaise provided the spark for her, and she would later encourage his fervor.

Unknown to the family, however, Curé Guillebert's theology had roots that were considered, in the context of post-Reformation France, borderline Puritanism. The Pascals' choice of a pastor had the effect of identifying them with a theological position that would come to be known as Jansenism. Jansenism was problematic in that it could be confused with a teaching of the Protestant reformer, John Calvin, on predestination. Predestination holds that God determines who will be saved and who will be damned, that our will is not free.

Curé Guillebert was a disciple of the French theologian Jean-Ambrose Duvergier, who was known by the title Abbé de Saint-Cyran. Saint-Cyran was imprisoned in 1638 by Richelieu because the king's conscience seems to have been afflicted by his confessor, Saint- Cyran. At the time of his arrest, Saint-Cyran had been the spiritual director of the monastery, Port Royal de Paris, for several years, and indeed the monastery under his spiritual direction was the center of his support. Saint-Cyran died in 1643. Pascal, who was a communicant at the Port Royal community, became involved in a major theological controversy. Saint-Cyran and Port Royal became the focus of a bitter theological dispute. This controversy is tangential to our purpose at this point, but it will arise later. The memory of this dispute and the rancor of the crackdown created animosities that influenced the course of French history.

Back to Paris

The Pascal family's association with Jansenism occurred concurrently with a new phase in their life. Etienne retired in 1648 and returned to Paris. By that time Blaise and Jacqueline were both adults and were attending Mass regularly at Port Royal de Paris. Jacqueline had decided to join the community of religious women at Port Royal. Blaise supported her decision at the time, but Etienne did not want to be without her. Jacqueline deferred to her father and agreed to postpone her entry into the community.

Shortly after the Pascals settled in Paris, a succession of politically motivated skirmishes, known as the Fronde, erupted, making life in the city dangerous. The family moved to Clermont in May 1649, staying with Gilberte and Florin Périer for more than a year. In retrospect, the Fronde was inept attempts to undo Richelieu's concentration of power in the monarchy. While easily neutralized at the time, the Fronde was energized by the same political and economic stresses that would destroy the monarchy in the next century.

When the family returned to Paris late in 1650, Etienne was a sick man. He died within a year, on September 24, 1651. Blaise was twenty-eight and Jacqueline twenty-six. Blaise had received his education and his formation personally from his father and Jacqueline had always been with him. They had never left their father's home. They had lost their beacon.

But Jacqueline knew what she was going to do. She was free to join the community at Port Royal. Blaise, whose permission was required, wanted her to relent. The prospect of losing his beloved father and closest confidant in so short a time was very difficult. Nevertheless, Jacqueline joined the Port Royal community early in 1652.

Blaise had a falling-out with her over their inheritance and over a dowry for Port Royal. Reluctantly, in May he agreed to her profession and a year later to the dowry. His correspondence with his sister, Gilberte, indicates a deep sorrow and a pessimistic focus on past spiritual failures. He was now alone, facing a period of mourning and spiritual discernment.

Pascal was never in robust health. He was subject to severe headaches, toothaches, and digestive upsets sporadically but often. The medical treatments were varied, intrusive, and only marginally helpful. The sorrow over Etienne's death was exacerbated by renewed illness, which must have contributed to his darkness. Despite his religious dispositions, he knew he had to find his way in the world. He set out to expand his social circle, make new friends, and use the gifts God gave him lest, without diversion, he would wallow in self-absorption.

He developed a new friendship with a former neighbor in Paris, Arthus Gouffier, Le Duc de Roannez, royal governor of the province of Poitu. Roannez shared Pascal's enthusiasm for the post-Reformation surge in the church's pastoral work for the poor, for the missions, and for the spiritual growth of its members. Roannez himself was in a period of reassessment in the face of an emerging political system as well as the fruits of the

Counter-Reformation. As a businessman, he was also interested in Pascal's efforts to use his scientific discoveries for profit. He empathized with problems in marketing his arithmetic machine.

Pascal spent the last four months of 1653 in Poitu with Roannez and friends, two of whom were avid gamblers. They expected little or no interest from Pascal in their strategic discussions. But Blaise lit a fire under them. He refined their gamesmanship by solving several problems they posed regarding the distribution of stakes in specified situations in dice games. Their discussions aroused Pascal's interest in the mathematics of probabilities where outcomes are equally likely (usual in games of chance).

Pascal soon established the systematics of probability theory. One outcome was the so-called arithmetic triangle, which is the basis for the binomial probability distribution, now a basic component in statistics. It describes the probability distribution of specific two-state outcomes, e.g., heads/tails in successive coin tosses. After the vacation, he continued his work on probability and compared notes on the subject with Pierre de Fermat, who was himself a well-known physicist and mathematician in Toulouse.

In addition, at this time Pascal wrote treatises on the *Equilibrium of Liquids* and on the *Weight of the Air Mass*. In this landmark work, he completed the analysis of the experiments done on the vacuum and on the relationship of the volume of a gas to pressure, the *Great Experiment*, executed by Florin Périer in 1648. On his foray into "the world," Pascal, while recharging his spirit, set the foundation for modern statistics and demonstrated the role of atmospheric pressure in creating nature's "abhorrence of a vacuum."

The Second Conversion
When Blaise returned to Paris in early 1654, he reconciled with Jacqueline, now Sister Jacqueline de Euphémie. When he visited her at Port Royal, he was also able to see Gilberte's two daughters, who were at the school there. Pascal had become recollected spiritually. He had come to realize that the promptings of the heart are always present and that the voice of God can be

heard if the noise of the secular world can be stilled. He knew that the lure of the senses muffles the whispers of the hidden God to the heart.

He laments that the world's gratification attracts more strongly but does not endure, while the whispers, though weaker, never go away. To achieve happiness that endures, the secular noise has to be turned off, so one can hear the eternal voice whispering to the spirit. The secular babble is always going to be around, but happiness lies in heeding the whisper.

In terms of the human will, (1) the eternal voice of the spirit is the greater lasting good, and (2) we cannot lose this voice unless we will to do so. We need to (1) develop the habit of shutting out the noise, (2) listening to the whispers of God, and (3) acting on the promptings of the heart. These are the habits of prayer—recollection, contemplation, and discernment.

In September 1654, Blaise—whose income was limited—moved to simpler quarters on the left bank, a short walk from Port Royal. Surely the location close to Jacqueline and the Périer nieces sweetened the move. His conversations with Jacqueline were frequent. He bared his soul, expressing contempt for things of the world and a lack of enthusiasm for intellectual accomplishments, for which he was becoming so well-known. He recognized the possibility of a deeper spiritual life, but he did not feel that God was offering it to him. He was undergoing a spiritual dryness—a dark night of the soul.

We know how God answered his distress now. After his death, two written pages were found sewn into the lining of his clothing, one on paper hastily scribbled, the other on parchment carefully executed. Each contained essentially the same information. This text, called his memorial, is as follows:

> The year of grace 1654
> Monday, 23 November, feast of Saint Clement
> Pope and martyr and of others in the Roman martyrology
> Eve of Saint Chrysogonus, martyr and others.
> From about half past ten in the evening
> until about half past midnight

FIRE

God of Abraham, God of Isaac, God of Jacob, not
of the philosophers and scholars
Certainty, certainty, heartfelt, joy, peace
The God of Jesus Christ . . .
My God and your God
Your God will be my God.
Forgetfulness of the world and of every thing except God
One finds oneself only by way of the directions taught in the gospel.
The grandeur of the human soul
Just Father, the world has not known you, but I have known you.
Joy, joy, joy, tears of joy . . .
I have separated myself from Him . . .
They have abandoned me, the fountain of living water
My God, will you leave me?
May I not be separated from Him eternally.
And this is life eternal, that they might know thee, the only true God
And Jesus Christ whom thou hast sent
Jesus Christ . . .
Jesus Christ . . .
I have separated myself from Him. I have run away from Him,
renounced Him, crucified Him.
May I never be separated from Him . . .
One preserves oneself only by way of the lessons taught in the gospel.
Renunciation total and sweet.[6]

The detail of this "memorial" bespeaks the observational habits of a scientist—clear, detailed, factual, uncluttered—complete in all that is temporal,

---

[6] O'Connell, *Blaise Pascal: Reasons of the Heart*

but it is a profound, passionate, spiritual record of a loving encounter with God. It consists of expostulations on the inner presence Blaise was experiencing, personal sorrow for his own failures and biblical references evoked by his spiritual vision—for example, fire: Moses seeing God in the burning bush.

This experience had to have mitigated his agony over priorities and guided him in his future choices. For the rest of his life, he kept this text sewn into his clothing. He lived actively and fruitfully from then on, but he clearly wanted this memory to guide the rest of his life. Indeed, as he neared his premature passing, he lavished charitable care on the poor, both in giving of his time and his substance.

From what is known of his life after this second conversion, there is no indication that he had shared his experience with anyone. It is believed that he made a retreat at Port Royal des Champs[7] in January 1655. His published interview with M. de Saci on Epictetus and Montaigne is attributed to this time. De Saci was the spiritual director of a group of laymen called Solitairies at Port Royal des Champs. De Saci's birth name was Isaac LeMaître, the son of the widowed Catherine LeMaître (née Arnauld), a nun at Port Royal, one of many Arnaulds associated with the monastery.

Pascal's association with the monastery at Port Royal would soon involve him in a historic theological controversy. The controversy is mentioned previously in the discussion of Abbé Saint-Cyran. In defense of his friends, particularly the priest-theologian Antoine Arnauld, he published eighteen tracts, *The Provincial Letters*, over fifteen months (January 1656 to March 1657), which are masterpieces of the polemic art, scorning the arguments of Jesuit opponents of Port Royal theology. These essays are French literary classics. But this controversy is tangential to our primary interest in Pascal's science and his personal religious experience.

In March 1656 Pascal's ten-year-old niece, Marguerite Périer, a boarding student at Port Royal, was cured of a serious and chronic eye infection by

---

[7] The original home of the Port Royal monastery.

the application of what was said to be a relic of Christ's crucifixion, the Holy Thorn. This event occurred at the time when the battle with the Jesuits was in full swing. Church authorities investigated and verified that the cure appeared legitimate. The event appeared fortuitous to the Port Royal community and, whether because of it or not, Pascal began to prepare a treatise on miracles. But as it progressed, it became a comprehensive apologia for the Christian religion.

His preparation is documented by voluminous notes of varying length—from mere phrases and sentences to lengthy essays. He organized them into broad areas, obviously in preparation for writing. These fragments, known as the *Pensées*, are the best known of his literary work, and were discovered after his death. In the spring of 1658, he is known to have had several extensive conversations with friends at Port Royal regarding his plans to craft a unique argument for the Catholic faith.

From these *Pensées*, Pascal argues that the true religion must address and reconcile two contradictory truths that color man's experience of himself—his greatness as well as his wretchedness. The true religion must integrate man's dark sense of himself with his utopian striving, his apparent brokenness as a being of contraries.

Saint Paul, in his Epistle to the Romans (Ch. 7:18–19) describes this wretchedness in the midst of his striving for righteousness: "I know that the good does not live in me, that is, in my human nature. For even though the desire to do good is in me, I am not able to do it. I don't do the good I want to do; instead, I do the evil that I do not want to do."

Pascal recognized that the philosophical proofs for the existence of God, while valid, appeal to the mind only but not to the heart. They do not attract the will. They lack rhetorical compulsion. Pascal's epistemology, his standard of truth, nested in the heart—what one knows interiorly because he is alive. We are not an empty slate.

He did not live long enough to complete the apologia, but his *Pensées* (literally "Thoughts") reveal a great deal about his thinking and particularly his broader understanding of what it means to prove something. The *Pensées*, a gold mine of provocative thought on a vast range of topics, should be consulted often but in small bites.

One example is Pascal's insight into what may or may not compel the human will is the last line of *Pensée* 345: "There is enough light for those who wish only to see and enough darkness for those of contrary disposition." The core insight is that God is a hidden God. He may be revealed to those who sincerely seek Him, but those who lack the desire to see will not see.

Pascal continued his intellectual output through 1658, writing theses on religious themes and on geometry, particularly a treatise on the cycloid, a curve traced out by the motion of a point on the circumference of a circle as it rolls in a straight line on a flat plane. In publishing his solution, he offered a cash reward for any mathematician who could come up with a more cogent analysis of the problem. But it was becoming clear to family and friends that his health was deteriorating.

His sister, Jacqueline de Saint Euphémie, died at Port Royal in the fall of 1661 at the age of thirty-six. In Pascal's final years, he devoted time and resources to the care of the poor. He shared his home with a destitute family, and when one of their children developed smallpox, he left his own home to them and moved to his sister Gilberte Périer's family home in Paris.

Early in 1662, he and Roannez established an omnibus service for the poor in Paris, the precursor of public transit. The demand was so great that very soon within the year, they had to augment the schedule.

Blaise Pascal died on August 19, 1662, at the age of thirty-nine. He is buried in the parish church, Saint Etienne du Mont, on the left bank. Blaise Pascal to this day inspires scholarship. He was a premier scientist, a mathematical innovator, the philosopher of the heart, an eloquent writer, a historic polemicist, the inventor of a prototype computer, and a public transit pioneer. The scholarship on Pascal has never let up. He was present at a pivotal time and place in the evolution of the modern world. He foresaw the weakness of the Enlightenment's presumptions and is a source of perspective for the postmodern muddle we are in today.

Alexis Carrel (June 28, 1873–November 5, 1944)

Alexis Carrel, a 1912 Nobel laureate in physiology and medicine, vastly expanded the experimental capacities in the life sciences as well as the surgical capacities in medicine. Over forty years, primarily at the Rockefeller Institute for Medical Research in New York City, he refined and expanded the art of *in vitro* tissue culture and the study of living organs. He investigated the organization and metabolism of living things on all levels of complexity, from cells to organs.

He laid the foundations upon which much of modern surgery has been built. With a continuous program of exploratory experiments and daring surgical trials, he evolved procedures that enabled surgeons to treat diseased or wounded organs, perform organ transplants, and indeed use machines to substitute for organs in surgery. His findings led to innovations such as blood transfusions, kidney dialysis, and open-heart surgery. Carrel's eclectic scientific initiatives made him a public icon at the peak of his career.

He was born in Lyons on June 28, 1873, the eldest of three children of Alexis Carrel-Billiard, a textile manufacturer, and Anne Marie Ricard. His grandfather had also been in the textile business as a linen merchant. When his father died, Alexis was just five years old. His mother raised him and a younger brother and sister. He was educated in local Jesuit schools up through college. An uncle had introduced him, as a youngster, to basic experiments in chemistry. As a biology student he developed a special skill in dissection of birds and graduated with an undergraduate degree in both letters and science.

At seventeen, in 1890, he enrolled in medical school at Lyons, arriving with a head start in anatomy and with an unusual manual dexterity developed in the practice of stitchery, a skill associated with the family business.

After completing the formal courses of study, he spent the better part of eight years (1893–1900) in practical training, mostly in local hospitals. He had a unique perspective on living creatures, perceiving them holistically—observing the parts in terms of the whole and speculating on relationships within the anatomical structure.

His special aptitude for surgical innovation found its outlet in 1898 in the laboratory of Dr. J. L. Testut, a renowned anatomist.[8] He also spent a year in military service as a surgeon in the French Army. Carrel was not interested in the routine procedures of medical practice—the diagnosis of disease and the treatment of the ill. He had no intention of setting up a traditional practice. He wanted to explore the most basic problems of medicine from the ground up. He may not have been able to articulate the scope of his interests initially, but in his career he explored the morphology of life, the processes of metabolism, the body's adaptation to pathology, and how living materials, from tissues to organs, operate.

Vascular Surgery

The medical school at Lyons was among the most prestigious in France. Carrel made almost no attempt to involve himself within the social environment, nor did he seek a mentor to help negotiate the formal requirements of various certifications. He pursued his research in vascular surgery while others in his class were concentrating on the professional licensing and certification examinations. His neglect of the protocols of passage marked him as a maverick. Not surprisingly, he failed to pass some of the certification examinations and to meet hospital association protocols.

In 1894 President Carnot of France had died at the hands of an assassin because a major blood vessel could not be repaired. This shortcoming of medical technique provided Dr. Carrel with a scientific problem to address

---

[8] *Dictionary of Scientific Biography,* Vol. III, Editor, C.C. Guillispie, Chas. Scribner & Sons, N.Y. 1971, pp. 90-91

when he graduated in 1900. From military service he had been trained in the treatment of wounds, which experience he generalized into an interest in how the body heals. He chose to tackle the challenge of repairing severed blood vessels.

To solve the problem, *first*, blood clotting had to be prevented. Clotting occurs when blood is exposed to air. Somehow, the bleeding had to be minimized. *Second*, the path of blood flow had to be isolated from the air. *Third*, infection had to be prevented, i.e., the procedure had to be aseptic, free of microbes. Carrel had to (1) isolate the main blood flow from the point of connection, (2) achieve the surgical connection with minimal bleeding, and (3) minimize the chance of infection.

Carrel was able to stretch and fold out the tissue of each severed end—much as you do when you curl up your sleeves. He flared out the tissue at each end to form a flat, flange-like ring at each rim, so that the ends of each vessel looked like the ends of bugles. He then butted the two flat, flange-like rims of tissue together and stitched them away from the blood passageway. The stitches connected the flattened rims of tissue, which were outside the blood passageway. He directed the needle and ligature parallel to the blood flow. When the coupling was sewn firmly, the mated flanges of tissue formed a band of flesh surrounding the blood vessel. The wound would heal outside the blood passageway, while the blood flowed inside, away from the connection site.

He designed very thin and very sharp curved needles to execute the stitches. The curvature of the needles made them easier to direct—to, through, and away from the narrow mated rims of tissue. He coated the needle and the ligature with a film of paraffin jelly, which covered the stitch holes in the tissue as the stitches were made. In addition, he carefully specified aseptic procedures in performing the work. The needles made very fine holes and the jelly coating made infection unlikely. The minimal clotting occurred outside the passageway on the rim. Carrel published this work in 1902 at the age of twenty-nine. Ten years later this work gained him the Nobel Prize.

Rejection and Fulfillment

For two years after the publication of this work, Carrel sought a research position in France suitable to his surgical talents, but there was no place for him. He spent the winter of 1903–04 in Paris in advanced study, and in May 1904—having found no appropriate position—he sailed for Montreal with a letter of introduction from his uncle, the Catholic bishop of Clermont-Ferrand.[9] In Montreal his leads did not work out initially, but he soon found a hospital position and, by virtue of his reputation, eventually landed an assistantship at the University of Chicago, where he resumed his experimentation in surgical techniques.

Soon his reputation as a surgical innovator came to the attention of Simon Flexner, who was setting up and staffing the Rockefeller Institute for Medical Research in New York. Flexner was an innovator himself. He was looking for imaginative researchers and bold thinkers who could extend the capacities of medical science in new directions. In 1906 Flexner brought Carrel to the Rockefeller Institute. Carrel had found his niche. He would spend thirty-three years at Rockefeller.

Angst at Lourdes

Carrel, from his writing and personal diaries, was committed to the rationalist intellectual orthodoxy of his time. He had adopted the philosophical mindset of post-revolution France in terms of science but, as we will see, with reservations. He did not endorse the attendant prejudice of rationalistic scientists who denigrated faith or any spiritual insight. He drew criticism for his willingness to consider such sources of knowledge.

When Carrel was a very new doctor, it was widely claimed that cures were occurring by virtue of prayer at a shrine of the Blessed Virgin Mary at

---

[9] Carrel, Alexis, *The Voyage to Lourdes,* 2nd Ed. Real-View Books, 2007, p. 25

Lourdes, an isolated town in the foothills of the Pyrenees in southwestern France. In this remote mountain setting, the Blessed Virgin was said to have appeared on eighteen occasions to a simple illiterate teenager, Bernadette Soubirous, in 1858. The visions took place at a point along a local stream, called the Massabielle grotto. The cures were associated with water that rose from a spring near the grotto.

On May 25, 1902, Carrel set out to investigate the reported cures. He agreed to accompany sick pilgrims on a train from Lyons to Lourdes. Carrel wrote a detailed personal account of the events at Lourdes shortly after the event. After his death, his wife, Anne, came across this third-person account and published it in 1950 as *The Voyage to Lourdes*[8] with a preface by Charles Lindbergh.[10] Names were changed, but it was clearly autobiographical. The book was reprinted in 2007. In a new introduction, Stanley L. Jaki presented details of Carrel's life since the 1902 trip, especially details relating to Lourdes.[11]

Carrel's account details how from May 25 through May 29 he was directly involved in events surrounding the cure of Marie-Louise Bailly, a twenty-three-year-old woman who was suffering from advanced tubercular peritonitis, an infection of the abdomen. He had seen her three times and had examined her twice prior to her going to the baths and the grotto. He confirmed the diagnosis and expected her to die soon, promising to become a monk if she recovered.

On the following day, when Marie-Louise was brought to the baths and the grotto, Carrel and several associates accompanied the caravan of sick people and their attendants to the baths. Marie-Louise was so weak that the attendants did not submerge her in the bath but poured a little water over her abdomen. After the visit to the baths, she and other sick people were taken to a service adjacent to the grotto. During the prayers, Carrel thought he noticed a slight change in Marie-Louise. Several minutes later

---

[10] Carrel, Alexis, *The Voyage to Lourdes,* Harper and Brothers, 1950
[11] Carrel, Alexis, *The Voyage to Lourdes,* 2nd Ed., Real-View Books, 2007

he was sure she looked better. His incredulity was deepened as the improvement appeared to progress steadily. He was startled at the dramatic speed of her apparent recovery.

After Marie-Louise returned to the hospital, he examined her and found her totally sound. Other physicians confirmed her recovery. He spent the evening discussing the event with an old schoolmate from Lyons who was a regular volunteer at Lourdes and to whom he had made the rash promise to become a monk. The next day he examined Marie-Louise again and found she was progressing well after a good night's sleep. Again, other doctors confirmed his findings.

Carrel maintained his belief, although with some difficulty, that there must be a scientific basis for the cure—there was no miracle. The most common medical explanation for what appears to be a sudden cure is hysteria, in which a strong positive emotion such as euphoria triggers the recovery from a physical infirmity, which infirmity had been due to the disordered emotional state in the first place. In other words, the infirmity is due to a toxic emotional state and when the emotional state changes, the infirmity disappears.

But in Marie-Louise there was no evidence of hysteria. She appeared emotionally balanced before and after the cure. Her answers to questions were focused and factual. She complained of being weak and tired, but she knew she had been cured. She had traveled to Lourdes hoping to be cured. She accepted that she had been cured without affectation of any kind. After regaining strength, she traveled back to Lyons by train, apparently alone. Dr. Carrel maintained contact with the case until Marie-Louise died thirty-five years later. While her blood tests did show evidence of tubercular infection, she never relapsed, living the very demanding life of a nun, ministering to the poor.

Carrel fully cooperated reluctantly but honestly in documenting the events he had witnessed. He knew his involvement would compromise his scientific reputation, but he recorded what he had seen. While he testified to what he saw, he was not willing to give up the search for a natural cause. In fact, he drew heat from both sides. The medical staff at Lourdes did not like his criticism of their facilities and methods, and he suffered the scorn of the medical profession for taking the study of "miracles" seriously.

At the onset of World War I, Carrel returned to the French Army as a surgeon. He and Anne de la Motte de Meyrie, a surgical nurse, served together treating the wounded near the front. During his service, Carrel and English chemist Henry Dakin developed a new procedure to sterilize very deep wounds. The Carrel-Dakin procedure accelerated the healing process and significantly increased the survival rate of wounded. It limited the long-term effects of the wounds and enabled more wounded to return to duty. The procedure became a standard of the healing art for some time until antibiotics were developed.

When Carrel returned to New York, he continued his research into keeping living tissue thriving outside the host organism, *in vitro*. He progressed to tackle the same problem with living units of higher complexity, eventually trying to make whole organs live and function outside their host organisms. His early work in growing cells and tissue *in vitro* was basic to his interest in studying living forms broadly, from their most fundamental manifestations up to living in a living host. In the process of this multilayered study of life, he perfected procedures that enabled the broader science of physiology to blossom. His techniques, in other hands, enabled major improvements in the way life could be studied. His later work in the physiology of organs introduced new techniques that readily transferred not only to other researchers but, more telling, to practicing surgeons.

Simon Flexner, who hired Carrel, became director of the Rockefeller Institute in 1920. At a testimonial dinner for Dr. Carrel in 1936, Flexner recognized the fundamental nature of Carrel's creative experimental techniques: "It may be said that Dr. Carrel has enriched surgery, physiology, and pathology through his scientific discoveries and is to be regarded as a benefactor of mankind."[12]

EARLY RELIGIOUS AMBIGUITY

Dr. Carrel had maintained several diaries and spiritual meditations in French throughout his life. When Mme. Carrel prepared his papers for donation to the archives at Georgetown University in 1949, she produced a French typescript of these writings that, along with the various originals, is available at Georgetown University in Washington DC.[13] Carrel's biographer, Joseph T. Durkin,[14] translated Carrel's diary entries from the French typescript.

During his years in medical training in Lyons and the subsequent period of uncertainty over finding a permanent position (1890–1904), Carrel was anxious over many things that which go to the heart of his motivation, ability, and self-confidence. The subjects of his anxiety are diffuse—no one citation is central. Basically, he was anxious and uncertain, as are many young professionals. A brief sampling of diary entries from the years 1896–97 from Durkin exemplify Carrel's doubts in the midst of his training.[15]

On his ability to know anything: "It is impossible really to know, at each step one is faced with mysteries; all is vague, obscure and inconsistent. In science, in physiology, in religion, one digs, but finds nothing."

On the wretchedness of his life: "Is life worth living? Why should I labor well, since my work will either avail me nothing or bring to me some satisfactions of vanity or money which, after a few years, fatigue or sickness or old age will prevent me from enjoying?"

---

[12] Durkin, Joseph T., *Hope for Our Time, Alexis Carrel on Man and Society,* Harper & Row, N.Y. 1965, p. xii

[13] Alexis Carrel Papers, Georgetown University Library, Special Collections Division.

[14] Durkin, *Hope for Our Time.*

[15] Durkin, *Hope for Our Time* pp. 55-57

On the selfishness of his motives: "I have been learning a science that can conduct me to an important position in the world, procure for me scientific success. . . . But, in the last analysis, I work actually not with a charitable or philanthropic intention. I want for myself intellectual satisfactions and a flattering of my self-love."

These doubts seem to be those of a young man anxious about the future and ultimate questions. However, what is also revealed in the transcript is Carrel's determination not to be dissuaded by doubts: "It is the need of a reasonable being that he believe in the rationality of the reasons for which he acts."[16]

He was struggling with the meaning of life and the way to truth. He was certain of God's existence but agonized over God's hiddenness and the difficulty of proving his existence. He saw human striving as futile in the face of the elusiveness of truth. He saw a career, no matter how successful, as ultimately meaningless. At other times he was motivated to be a success in life but brooded over the selfishness of his motives. Would he be happy? What could make him happy? Regarding his profession, he was not taken with a career in medicine, and references to marriage suggest a lack of enthusiasm consistent with his fatalism about life in general.

Carrel believed in God and prayed, thought about, and meditated on religious matters. He recognized the reality of the spiritual life in man, but from a distance. He preferred to gather scientific evidence on the phenomenon of faith rather than simply open his heart to love, the language of God to the heart.

Carrel's Unique Position

Alexis Carrel had attained his position at the Rockefeller Institute in 1906 because Simon Flexner was convinced that his unique approach to science

---

[16] Durkin, p. 56

was just what the Institute needed to fulfill the innovative mission he foresaw. The Institute had a basic commitment to medical research, which made it an ideal place for him—perhaps the only place. He was able to follow his own instincts and desires to a degree that was rare. He thought and operated boldly on a creative plane far higher than others. A unique professional, he had no peers. Carrel's grating manners were of no consequence during his peak years at the Institute. And as long as Flexner was in charge, nothing was going to change. It was also clear that Flexner's choice had been brilliant.

Carrel's unique talent had a downside. The profession had no experience of such a nontraditional and extensive research strategy. His colleagues viewed his public visibility as self-promotion and unprofessional. Maybe Carrel could have engaged his colleagues if he were not a loner. It would have been fruitless for him to look for his colleagues' endorsements.

When Flexner retired and the management of the Rockefeller Institute passed on to Dr. Herbert Gusser, a corporate culture emerged. It became clear that no exception would be made for Carrel when he reached the retirement age of sixty-five. He was required to retire. In 1938 Carrel published his major work, *The Culture Organs,* coauthored with Charles Lindbergh. The book presented the results of their collaboration since 1930. Carrel retired from the Rockefeller Institute in the summer of 1938.

P. S. An Essay on Man

Know then thyself, presume not God to Scan;
The proper study of Mankind is Man.

Plac'd on this isthmus of middle state,
A being darkly wise, and rudely great
With too much knowledge for the Sceptic side,
With too much weakness for the Stoic's pride,

He hangs between; in doubt to act, or rest,
In doubt to deem himself a God, or Beast;
In doubt his Mind or Body to prefer,
Born but to die, and reas'ning but to err;

Alike in ignorance, his reason such,
Whether he thinks too little, or too much;
Chaos of Thought and Passion, all confus'd;
Still by himself abus'd or disabus'd;

Created half to rise, and half to fall;
Great lord of all things, yet prey to all;
Sole judge of Truth, in endless Error hurl'd:
The glory, jest, and riddle of the world!

Alexander Pope

Chapter 2

The Emergence of Modern Europe

THE BEGINNINGS OF SCIENCE
THE FEUDAL SYSTEM

*P**ax Romana*, the peace of Rome was the major factor in European history prior to the fall (ca. 475). The empire controlled the whole Mediterranean Sea from the Atlantic to Asia Minor. The Roman Army kept the "barbarians" at bay on the perimeter and insured the rule of law within. St. James evangelized Spain in the west while St. Thomas established the faith in Goa, India, in the east. The epistles of St. Paul give witness to the mobility possible within the empire. Wherever he traveled, he remained subject to Roman law. As a Roman citizen he had to be tried in Rome. Roman law required his extradition before he was martyred.

With the fall, "barbarian" tribes were free to plunder. Political and economic interplay contracted. Europeans had to improvise local systems of defense and governance to ensure order. The feudal system evolved, providing stability in local communities. Post-Roman Europe became a diffuse collection of agrarian societies built on local contracts between the land-owning nobles and the working-class serfs.

The Church, emerging from centuries of persecution, was a strong presence in every community and had a major role in administering the sacraments and supporting the afflicted. As a familiar and trusted presence, the

Church was a natural guarantor of justice between the nobility and the working class.

In addition, through its monasteries, the Church provided cultural and learning centers. The monasteries preserved ancient religious, historical, and intellectual records and absorbed intellectual influences from other cultures, particularly the Arabic.

As time went by, the barbarians were evangelized. The feudal system began to evolve. Seeking political and/or economic advantage, princes made alliances, changing the political map. Larger political units created a range of levels among the nobility.

When the societal base was no longer the local fief, the parish pastor lost some closeness with his flock and with his bishop. As political units became larger, the hierarchy incrementally but continually became a formal element in a more complex political and economic system, further isolating the pastor in his parish.

The Church hierarchy became a formal element in a complex political and economic system, a circumstance that undermined her pastoral role.

In Pascal's France, the pastoral role of the Church was conflated into a polity of posture and privilege. The local pastor, adjudicating disputes and administering charity, had faded away. Some "pastors" were able to accept benefices, i.e., titles with income from land. In some cases their responsibility to minister to the flock was optional.

Also, the culture of Europe was influenced in the tenth and eleventh centuries by the crusades, opening Europe to the Byzantine and Moslem cultures of the East. By the fourteenth century, the economic, political, and intellectual climate had been swept clear of medieval parochialism. But the privileged class endured in an outmoded system that had lost its original *raison d'être*.

By the sixteenth century, the age of discovery had opened the Far East and the Americas to European exploration and colonial exploitation. In addition, Luther's Reformation in 1517 had revealed a Catholic Church ill-suited to ministry within the culture.

By Pascal's day France had interests in trade and empire. Louis XIV was the epitome of absolute power. The Church had become entangled with the "things of Caesar" to the detriment of her pastoral mission. In addition, many commoners were uncommonly influential, wealthy, and powerful. The poor were still with them in an insensitive and unjust framework.

MEDIEVAL SCHOLARSHIP

Science, the experimental pursuit of knowledge of the physical world, began to take root in the millennium-old Christian culture of Europe in the fourteenth century. Embryonic scientists lived in a thousand-year culture that saw the physical universe as the work of a personal Creator in whose image they were made.

They knew the physical universe really existed, because God had created it and them. They saw order in the universe consistent with a supreme intelligence. They trusted that their senses did enable them to make measurements that could detect that order. God would not mislead them. As Einstein said centuries later, "God is strict, but He is not unkind."

The magnificence of the evening sky certainly enthralled our ancestors. Today, the light from our cities blurs the awesome spectacle of the firmament. Spending a night at sea when the moon is dark is an unforgettable experience. When science began in Europe, the motion in the heavens had been studied for centuries. Our ancestors had provided ample records of celestial motion, which drew the early scientists to astronomy especially.

The Beginnings of Scientific Thinking

Stanley L. Jaki, in his *Science and Creation*[17], shows that scientific thinking had begun to develop in many cultures before medieval Europe. But in each case the culture had failed to support scientific study. Approximately half of

---

[17] Jaki, Stanley L, *Science and Creation,* Scottish Academic Press, 1986. Fr. Stanley L. Jaki, a Benedictine priest, was recipient of the Templeton Prize for 1987

Jaki's seminal work is devoted to reviewing how every such beginning failed to nourish experimental study of the material world. In each beginning the gestation of the scientific method was compromised by some cultural impediment, such as superstition, for example, or astrology.

But Christian scholars such as Adelard of Bath (ca. 1125) had begun to recognize that the workings of the material world had a rational component that could be understood on its own terms. He asserted simply that "the realm of being is not a confused one, nor is it lacking in disposition, which as far as human knowledge can go, should be consulted."[18] Embryonic science existed in ancient Greek and medieval Muslim cultures. Why, then, did science only persist when it emerged in Europe between 1250 and 1650? Jaki argues that medieval Europe was the first culture where the widespread belief in a loving personal God, creator and maintainer of the universe, had endured for more than a thousand years. Because this culture had been built on belief in such a God, it had an innate assurance that the universe had to be ordered as well as the confidence to believe that persistence would detect that order. Note the implicit limit, "as far as human knowledge can go." This deep trust in our ability to know, within the limits of our contingency, was the Christian warrant for science.

This culture understood that the universe was a gift of the Creator and had to make some sense. The medieval scientists also trusted the intellectual tools they were given—gifts they had received from the loving Creator. They trusted that measurements were meaningful. Most importantly they had no doubt that the objects they studied really existed. In short, this culture was steeped in realism, anchored in a firm belief in a Creator God, who is both benevolent and trustworthy. Up to that time the primary influence on scientific thinking in medieval Europe was Aristotle's physics.

John Buridan, a scholar who lived in the early to mid fourteenth century, is cited by Jaki as an exemplar "of how the thinking of the medieval

---

[18] Ibid. p. 219

'physicists' was strongly influenced by their belief in a personal Creator." Buridan, based on his sense of the Creator and creation, rejected several restrictions in Aristotle's *Physics*. For example, Aristotle thought that matter in the world, sublunary matter, and the matter in the firmament, superlunary matter, were different. Aristotle's presumption was based on differences he perceived in the motion of the two systems. Buridan's insight was that space, time, matter, and motion were universal realities. The unitary nature of space, matter, and motion was the concept Newton used.[19]

Buridan is also important as a precursor of Galileo in that he proposed a concept of motion he called "impetus," which he described as an "impressed force" on a body.[20] To exemplify his idea, Buridan refers to the effect of a magnet on a body. He is also aware of the resistance to motion (friction). He describes impetus "to be measured by the velocity and the quantity of matter of the body in which it inheres." He has intuitions of mass, velocity, force, friction, and momentum, but he cannot sort them out. He has set the table for Copernicus and his followers, both in describing particle motion and in dispensing with the ideas of Aristotle when deemed incongruous with his insights.

The Role of Philosophy

The medieval stewardship of the monastic orders had preserved the ancient classics, which are primary resources for scholarship. Also, monasteries were repositories for the scholarship of their intervening medieval age. To call this era "the Dark Ages" is not descriptive. As a result of their stewardship, their philosophy had a strong theological cast. When the "new learning" began to flower in later centuries, the clerical scholastic philosophers were somewhat

---

[19] Ibid. p. 232

[20] Lindberg, David C., *The Beginnings of Western Science,* Univ. of Chicago Press, Chicago and London, 1992, p. 303-4. Lindberg gives a detailed description of Buridan's idea of impetus.

dismissive. Some acted as if they had sole proprietorship of classic philosophy. When they allowed Aristotle's physics to be conflated with Church teaching, a major and most unfortunate mistake was made.

What Is Philosophy?

The word *philosophy* comes from Greek and means "love of wisdom." Philosophy is the most general of human sciences.[21] It seeks to answer fundamental human questions, such as: What does it mean to exist? How do we come to know truth? How should I live my life?

In the tradition of Greek philosophy, metaphysics is the primary area of philosophy. Metaphysics means "beyond physics" in the sense of transcending physics. What underlies reality? What does it mean to be? The primary character of metaphysics resides in understanding what it means to exist. What is being as such? Metaphysics can reach into other areas of philosophy, providing a ground for epistemology, the philosophy of knowledge, and for ethics or moral philosophy.

Aristotelian metaphysics, a classical platform for developing philosophy, establishes five characteristics of being, one of which is knowability. If being is something that is knowable, then knowing being—i.e., knowing how something works—provides the knowing person with choices on how to use what he knows about being. Man's freedom, his ability to choose, depends upon his knowledge of being. Human culture is built upon a continual expansion of knowledge and the use of freedom in the application of that knowledge.

The person reveals himself by the quality of the choices he makes. It is quite reasonable to infer the character of a person from his choices. This critique is the basis for establishing a philosophy of behavior, ethics.

---

[21] Rizzo. Anthony, *The Science before Science: A Guide to Thinking in the 21st Century*, Press of the Institute for Advanced Physics, Baton Rouge, LA, 2004. This volume addresses the need for a sound philosophy to uncover the foundations of science.

Aristotle also wrote on physics, giving his analysis of the workings of the material world. His physics was not reliable because he was not able to make precise measurements. In the absence of experiments, he trusted appearances and was misled. Aristotle would applaud the heliocentric insight of Copernicus and the *Principia Mathematica* of Isaac Newton. Since the fifteenth century, the progress and influence of science have continued.

The Development of Scholasticism

Monasticism developed in the fourth century in the Christian East. By the early Middle Ages, monasticism had spread throughout Europe. Monks lived in, but not of, the world, practicing a discipline of work and prayer centered in monasteries. They were valued assets for their religious life and their intellectual and practical skills.

The first century fathers of the Church absorbed the ideas of Plato and his student, Aristotle. Later Christian philosophers, such as Augustine in the fifth century and Boethius in the sixth century, also passed on the Greek tradition. Christian monks also recorded the writings of later scholars such as the Moslim, Averröes, and Jewish Moses Maimonides in the twelfth century.

At the time of Copernicus, the major synthesis of classical philosophy was the thirteenth-century *Summa Theologica* of St. Thomas Aquinas. Aquinas, primarily a theologian, gave a systematic rendering and extension of Aristotle's work, called scholasticism. His *Summa* is a deliberate presentation of his adaptation of Aristotle's philosophy.

In light of the sources existing at the time, Aquinas's method is to pose a problem or question. He next presents every known position that has been taken on the problem, other than his own. In turn, he refutes each proposed position. Finally, he presents his own position and justifies it. The *Summa* of Aquinas stands on its own as a major philosophic work. Like other important works of philosophy, it is still a benchmark for development to this day.[22]

Philosophy as an Aid to Understanding

A particular benefit of scholastic philosophy as a discipline is its ability to ground being and truth, central underpinnings of science. But philosophy generally provides a guide for development of intellectual understanding. If we understand the philosophical grounds of reality and truth, we have a means of comparing, in a rational way, the structure of arguments presented in the exploration of an issue. This understanding can help to pinpoint the source of disagreement and provide an analytical tool to isolate the difference on some underlying issue.

Electrical engineers know that communications networks can fail in three ways. There can be a problem with the transmitter of the signal or with the receiver of the signal or with the medium that carries the signal. It is important that the sender and the receiver be engineered compatibly. In addition, the communication channel must be protected from noise. In personal communication, the compatibility of speaker and listener depends upon language, personal perceptions, shared human experience—how we understand reality, truth, beauty, etc., all elements basic to philosophy. Philosophy creates the framework that enables language to convey an idea.

Now an idea may be rejected even if accurately presented, clearly transmitted, and fully understood. In such a case the reason for the disagreement can often be identified by examining the philosophy of the communicants—their ideas on reality, truth, and beauty, for example. Mutual philosophical sensitivity may enable the participants to identify and examine reasons for their disagreement. Philosophy affords a capacity to look at such issues on a basic level.

---

[22] For a contemporary adaptation of Thomism, see W. Norris Clarke's *The One and the Many: A Contemporary Thomistic Metaphysics*, U. of Notre Dame, 2001

The obstacle to compatibility is often a cultural divide between the speaker and the listener. An idea may be rejected because it is incompatible with the listener's habitual commitments, how he has set his sails in life, his *modus vivendi*, manner of living. When attempting to frame a religious belief, it is prudent to recognize the personal commitments of the listener. His worldview may be the basis for a rejection. Such entrenched differences are analogous to noise in an electronic communication channel. In such cases, philosophy can often be an analytical means to achieve understanding while continuing to disagree.

We will try to show how the senses, the mind, and the human heart are means by which we know. Paralleling these means are perception on the order of the senses, reason or the reach of the mind, and love or the order of the heart.

The New Philosophy of René Descartes

René Descartes (1596–1650) was a mathematician and philosopher who belonged to the Mersenne Academy[23] with Étienne Pascal, among others. He was interested in problems in mechanics and is responsible for the common technique of graphing how one variable quantity varies with another. How, for example, velocity varies with time—the Cartesian coordinate system.

While on a visit to Paris in September 1647, Descartes, fifty-one at the time, met with Blaise Pascal, twenty-four, for two days. Pascal had moved to Paris from Rouen to seek medical attention for a recurring illness. He apparently was ailing during the visit. They discussed his calculating machine and his experiments on the vacuum on the first day, with some others attending. On the second day, they conferred alone. No one knows what they discussed, but it is surmised that Descartes, who had some knowledge of medicine, may have discussed remedies for Pascal's illness.[24]

---

[23] The Mersenne Academy advised the king on the current state of science.

[24] O'Connell, p. 29

The Philosophical Logjam

When Descartes turned his hand to philosophy, he produced a system that was broadly adopted for modern scientific understanding. His epistemology has had a profound effect on modern thinking. His objective was to create a totally new starting point. Classical philosophy, or scholasticism, was viewed for a millennium as the domain of the medieval schoolmen.

Descartes, a Catholic Christian, was sincerely motivated by the perception that scholasticism was dated. It lacked fecundity. It existed in a theological milieu that gave it a plodding and formal caste. The "new learners" disliked its ritualistic origins.

Descartes wanted to jump-start philosophy. Renewal required a beacon of certainty—something everyone could see and find convincing, obvious, and undeniable. He wrote extensively in his *Meditations* and in his *Discourse on Method* on the new rules he perceived to constitute valid procedures in argument. In his work, predominantly in the first-person singular, he proclaimed himself unbiased and interested in no particular end. He was hoping to break what he saw as a logjam of contention among philosophers. Although they acted in good faith, they disagreed on many things.

In his 1637 *Discourse on Method,* he ruminates at length on why philosophers of competence and good-will fail to agree on so many things. He speculates that the savants might reason more consistently if a new starting principle could be found—a principle that would be impossible to doubt. His reflections led him to the proposition that he could not doubt his own existence because he was aware of himself thinking. In Latin, this proposition is, "*Cogito, ergo sum*"; in English, "*I think, therefore I am.*"

While there are problems with this starting point, we have to recognize two things. First, and most important, Descartes's *Cogito* succeeded in marking the trail for modern philosophy. In the spirit of the Renaissance, Descartes provided the stimulus from which modern philosophy progressed. His initiative caught on, filling a perceived need. Spinoza and Leibniz, who followed him, reacted to his system, and the evolution of modern philosophy began.

Second, although scientists may or may not have thought through the implications of the *"Cogito, ergo sum"* proposition, there was a general acceptance of the new philosophy. Scientific experimentation from that time on followed the Cartesian lead. Parenthetically, Pascal did not accept the *Cogito* philosophy.

Reductionism

Descartes's philosophy suggests that the scientist is an isolated subject, an observer, standing in space before an object with absolutely no effect on the object. What exists is an isolated subject, the scientist, separated in space from the object and not affecting the observed object in any way.

This posture allows the scientist (subject) to characterize the object by properties identified in the mind of the subject, himself. The reality of the object is circumscribed by the abstraction of the observer. There is no recognition of the objective essence of the object. There is no consideration of the object's being, as such. There is blindness to the ontological nature of the object, its status as an objective reality, a real being.

The view is ontologically licentious, because the interest in the object is completely subjective, failing to infer any essence to the object itself. The scientist considers a reduced set of characteristics of the object, a limited set of measurable properties. There is a philosophical consequence to this reduction. The value of the analysis can be vitiated when the reduction significantly diminishes the essence of the object. What is clear about this model is that it makes the "idea" of the object within the mind of the observer the item of interest. There is no interest in any other aspect of the existing object. It beggars the object, reducing it to ontological poverty. The reality of the object is suppressed. Realism is out the window.

Deep in Idealism

Descartes's philosophy of science identifies the object of an experiment with the idea the scientist forms of it. The being of the object itself is circumscribed, ignored. The scientist selectively abstracts from the object the characteristics of interest to him. The grasp of being, in such an understanding, is reduced *ad hoc* to the physically measurable. Being is isolated from the reach of the mind and the promptings of the spirit.

However, each existing object is a fully independent being with a unique identity, whose fullness of being cannot be accessed by science. Science can access quantifiable characteristics, structures, processes, and the like, but its scope does not exhaust all that can be known. The scientist's idea of a "real object out there" is the *means* by which he knows the being, but his idea is not the being. The grounding of science, the identification of the independent existence of the object studied, is compromised as a matter of method. There is absolutely no need to force this nullity, this "no-thing-ness," on the object. This is the enduring problem for modern philosophy.

When the interest in an object is mundane, this oversight is of minimal consequence. But when some proclaim that the idea of the object is all that is or can be known, they are seriously overreaching. The effect in the scientific study of man, for example, is to ignore, as a matter of method, spiritual nature.[25] Today, this metaphysical overreach is showing up as some scientists recognize that physical theory becomes more remote and arcane.[26]

In the first paragraph of "Part Four" of his *Discourse on Method,* Descartes ruminates:[27]

---

[25] Kass, Leon R., *Life, Liberty and the Defense of Dignity: The Challenge for Bioethics,* Encounter Books, San. Francisco, 2002. See especially, Chapter 10, *The Permanent Limitations of Biology*

[26] See *The Idealist Trap* (below)

[27] Descartes, René, *Discourse on the Method,* Encyclopedia Britannica. 1952, p. 51.

I do not know that I ought to tell you of the first meditations there made by me, for they are so metaphysical and so unusual that they may perhaps not be acceptable to everyone. . . . I resolved to assume that everything that ever entered into my mind was no more true than the illusions of my dreams. But immediately afterward I noticed that whilst I thus wished to think all things false, it was absolutely essential that the "I," who thought this should be somewhat, and remarking that this truth, "I think, therefore I am," was so certain and so assured that all the most extravagant suppositions brought forward by the skeptics were incapable of shaking it, I came to the conclusion that I could receive it without scruple as the first principle of the Philosophy for which I was seeking.

However, Descartes fails to recognize that his resolution to doubt depends upon his lifelong interaction with the world. He uses language to announce his discovery. Did he not, from infancy, learn from innumerable stimuli from other beings? Descartes's ability to think was contingent on this endowment—the endowment of so many other beings and the language he fashioned using them.

His thinking was not, from the beginning, an existing capacity. He had to absorb languages, his tools for thinking, by interacting with the world within a culture. Acquiring a language requires copious interaction with real objects. Descartes, like the rest of us, had to realize his potential for thinking before he could do philosophy. The development of Descartes the thinker required incessant penetrating interaction with the "real world." What might have been more nearly correct was *"I am; therefore, I think."* The *"I am"* part is a long, contingent development that enabled his ontological endowment of intellect to be realized.

In learning the languages, he knew he had to root the nouns and the verbs in real beings or actions, the adjectives and adverbs in qualities he saw in real beings or actions, and the prepositions and the conjunctions in relationships he saw between or within real beings or actions. The subtleties of language rest on differentiations, discriminations, implications, inferences, presuppositions, insights, idioms, plays on words, etc. The art of poetry makes use of every modulation of reality in human experience.

We are born individually and arrive like a sailor adrift in a lifeboat. We are totally dependent on others and on the world itself. Eventually we realize we exist in a culture and in a universe of being. We mature as members of that culture, learning its customs, history, laws, and institutions. We have to be led through experience of the world in the context of culture before we can think scientifically. We have to know before we can think.

The idea we form, based on an experiment, is the *means* by which we grasp a limited quality of the reality of a being. The idea we form is not a circumscribed cradle within which we can exhaust reality. Scientists gain some knowledge, not all there is to know of a being.

The emergence of science in the fifteenth and sixteenth centuries was the work of realists—people whose millennium-old culture recognized a personal, loving Creator God who made and maintained all things. Their perception of the object of an experiment "standing against" them was real.[28] Their classical philosophy, since jettisoned, grounded science in realism. Nobody lives as an idealist in practice. The point has been made, pithily, by Etienne Gilson:[29]

> *The first step on the realist path is to recognize that one has always been a realist; the second is to recognize that, however hard one tries to think differently, one will never manage to; the third is to realize that those who claim they think*

---

[28] The Latin root of the word, "object", denotes something that "stands against" something else. The verb, "to object", retains the sense of opposition. The noun, "object" denotes a "real object out there."

[29] Gilson, Etienne, *Methodical Realism,* Christendom Press, 1990, *p. 127.*

differently, think as realists as soon as they forget to act a part. If one then asks why, one's conversion to realism is all but complete.

The Idealist Trap
Several physicists[30][31][32] looking at the state of physics today believe we may have made a mistake somewhere. Each of these writers has a different input on the problem. In effect, they suspect the grounding of physics.

When the object of study is simple and/or the purpose is limited, then reductionism can be practiced without consequence. The reductionist mind, however, has to be conscious of ignoring something. It may be inconsequential in most cases, but it is there. As science studies complex systems, such as living systems and psychic systems, this reductionist distortion has greater consequences. It becomes a trap for the idealist.

The trap is scientism, which asserts that science is the only means by which we can know anything. The belief implies that the scientist may reduce as he chooses. Scientism, a direct consequence of Cartesianism, dismisses the reality of the object and allows the unwary to assume there is nothing else to know.

Scientism is philosophical blindness, a disability, a refusal to see deeper or, more often, a refusal to look deeper. Scientism impoverishes the intellect, promotes relativism, and impedes rational discourse. It is the malady of debunkers.

The shortcomings of the Cartesian model do no harm as long as the reductionism and idealism are recognized for what they are. Failure to recognize the inherent idealist thrust of Descartes, and to take seriously the consequences for science, can lead to some serious errors.

---

[30] Laughlin, Robert B., *A Different Universe,* Basic Books, 2005

[31] Rosen, Steven, *The Self-Evolving Cosmos,* World Scientific Publishers, 2008

[32] Smolin, Lee, *The Trouble with Physics,* Houghton Mifflin, 2006

In the beginning of the scientific revolution, certainly up to and well beyond Newton's time, the philosophic distortion of the objects of scientific study had no consequences. But current physics may be showing the shortcomings of the idealism in theoretical physics.

THE SCIENTIFIC REVOLUTION

The study of motion, mechanics, identifies three characteristics relevant to motion—mass, length, and time. Mass is a measure of the amount of matter in something. Space and time are intuitive constructs. Science defines mass, length, and time operationally, by describing how to measure them. Science, at least in its beginnings, does not get involved in what space and time are, or what it means to exist, nor does it care about what truth is. Thomas Aquinas identified the "practical reason" as the primary interest of man. We want to know first what we have to do to get along in this world.

THE EMERGENCE OF CLASSICAL PHYSICS

Although Nicholas Copernicus (1473–1543), often cited as the initiator of modern science, was a Polish priest, clerics were not the sole practitioners of the "new learning." New techniques and competencies attracted skilled laymen, whose talents, perceptions, and experience brought freshness to science.

The sixteenth and seventeenth centuries brought a parade of breakthroughs, particularly in astronomy. International trade depends on navigation. England's historic colonial ascendancy depended on her navy. Astronomy was more than a casual interest of political and business leaders.

Copernicus upset an ancient philosophic premise when he proposed that the earth was not the center of the universe but that it revolved about the sun. Copernicus knew that this heliocentric model of the universe would anger theologians, who interpreted Genesis literally and also subscribed to

Aristotle's *Physics*. He anticipated the opposition of the schoolmen and dedicated his 1543 paper on planetary motion (*On the Revolution of the Heavenly Spheres*) to the Pope.[33] He softened his heliocentric posture as "mathematics written for mathematicians." This point of view made sense in analyzing the observed data. But it contradicted the ancient opinion of Aristotle that the earth was the center of the universe.

Johannes Kepler (1571–1630), a skilled mathematician, used the observations of planetary motion taken by Tycho Brahe (1546–1601) to formulate three mathematical properties of planetary orbits about the sun. The first of Kepler's laws is the simplest. It states that the path of any planet about the sun is an ellipse with the sun at one focus of the ellipse. The other two laws are quantitative and show (1) how the speed of the planet in orbit is related to its distance from the sun and (2) how the period of rotation (the year) varies with the mean radius of the orbit. This work had been published in 1609 and showed that the heliocentric model simplified the analysis of planetary motion. This model does not defer to prevailing philosophy, but it simplifies the problem.

Galileo Galilei (1564–1642), unlike Copernicus, was not overly concerned about the orthodoxies. His position on the center of the universe was a practical matter of how to report observations of celestial motion. He cared not about the implications of where he put the center of the universe.

Galileo, who was consistently improving the range of his telescope, had charted the details of planetary motion throughout his active lifetime. He knew, through Kepler's work, the laws of planetary motion, given in terms of how the planets moved relative to the sun. In terms of absolute certainty, his observations did not constitute a proof of the centrality of the sun. But they were strong indications that the earth might indeed orbit the sun also.

This situation exemplifies what Blaise Pascal refers to as the requirement that a proof be persuasive. To Pascal, proofs are arguments that persuade. A

---

[33] McMullen, Ernan, Editor, *The Church and Galileo,* U. of Notre Dame Press, Notre Dame, Ind., 2005. Michel-Pierre Lerner, *The Heliocentric "Heresy",* pp. 13-14

reasoned metaphysical argument may be completely and logically airtight, but if it does not have the impetus of persuasion, it withers. Galileo was persuaded by his observations.

Logical formalism may not have been met, but Copernicus's idea made perfectly good sense. The sun was the point relative to which the motion of the earth should be described, primarily because it makes analysis simpler. Galileo was not persuaded that biblical accounts were adequate determinants of celestial motion. For his boldness he was harassed by Church authorities.

In 1616 Galileo had had interactions with authorities in Rome, which he apparently sought initially. The result was a declaration that stated, in substance, that the heliocentric theory was both philosophically and theologically untenable. This statement was more formal and made significant distinctions regarding degree of theological error ranging from "formally heretical" to "erroneous in faith."[34]

There was no censure of Galileo then. Cardinal Robert Bellarmine, who delivered the message to Galileo, issued an auxiliary statement that Galileo was not abjured in Rome, i.e., forced to recant "on any opinion or doctrine of his." Bellarmine just delivered the message. The exculpation of Galileo apparently was necessary to dispel slander against him.

But by 1632 Cardinal Bellarmine had died, and Galileo was preparing to publish his *Dialogue on the Two Chief World Systems, Ptolemaic and Copernican* in March.[35] The book, written in Italian for widest circulation, was in the form of a dialogue on the merits of the geocentric and the heliocentric theories. At the time, Galileo, perhaps by design, was a bit heady in his dealings with Church authorities.

---

[34] Ibid. Annibale Fantoli, *The Disputed Injunction and its Role in the Galileo Trial*, pp.118-119

[35] Ibid. Michael H. Shank, *Setting the Stage, Galileo in Tuscany, the Veneto, and Rome*, pp. 78-7.

He was given two requirements by the censor. First, he was to print the text of a preface to the book, which he transcribed verbatim. But in addition he was told to write a final peroration endorsing a certain principle on divine omnipotence. The principle was a reference to an idea that the Pope, Urban VIII, had expounded in a recent encyclical.

Galileo complied with the request, but in a most inappropriate way. In the text of the *Dialogue* one of the characters who took part in the conversation was a dimwit named *Simplicio*. Galileo put the thought suggested by the Pope in the mouth of *Simplicio*. It was not well received.

Galileo was charged with violating his promise to present the two viewpoints in a hypothetical way without advocating either one. The charge analyzes the rhetorical means Galileo took to circumvent the objective stance he had promised.

On June 22, 1633, Galileo was found to be "vehemently suspect of heresy."[36] He was forced to abjure the heliocentric theory and signed a text of abjuration prepared for him. In his trial, no consideration of what he had observed was allowed, because it would not constitute a proof regarding the earth.

Galileo is honored for his courage in standing by what his data showed. In retrospect, his explanations were not as complete as they would later be, but he was fundamentally correct. Cosmology today affirms there is no center of the universe.

Galileo's work beyond the issue that drew his censure by the Church was varied and important. He made significant improvements in the optical telescopes of his day and was able to extend the range of telescopes to more distant planets.

He determined that the motion of a freely falling body was characterized by a constant acceleration. Acceleration is the rate of change of velocity

---

[36] Ibid. Francesco Beretta, Galileo, *Urban VIII, and the Prosecution of Natural Philosophers*, pp. 250-255

with time. He also showed that this acceleration was independent of the mass of the body, again contradicting Aristotle.

Galileo also addressed the problem of motion seen by different observers moving relative to each other. The water from a hose moves faster if you are approaching the hose than the water moves if you are moving away from the hose. This realization is described as Galilean relativity. Einstein developed his theories of special and general relativity, in terms that Galileo had invoked for the simpler situation.

From Galileo's day onward science continued to progress for four centuries, but he had forebears who saw beyond Aristotle by virtue of Christian cultural sensibility.

The censure of Galileo was bad philosophy and bad theology. It was an enormous black eye for the Catholic Church. Conflating Aristotle's physics with a literal reading of Genesis led them to reject a legitimate way to handle a problem in relative motion. There were no theological or moral issues involved.

Although Galileo was imprudent in flaunting his position, the schoolmen were isolated by a rash projection of their philosophy. In truth scientists from Buridan through Copernicus had challenged Aristotle's *Physics* for two centuries. The schoolmen were hanging on to Aristotle while laymen, with a Christian cultural confidence, were clarifying our understanding of God's creation. They might well have heeded the wisdom often posted on laboratory walls: "If you think it can't be done, get out of the way of the one who is doing it."

Isaac Newton's Physics and Mathematics

Kepler's discoveries gave Isaac Newton (1642–1727) quantitative clues that helped him determine his law of universal gravitation. Gravitation is the name given to the type of force that the sun exerts on the earth and that the earth exerts on us. A gravitational force acts between any two objects in the universe by virtue of their mass.

The law of universal gravitation describes how the force of gravity depends upon the mass of the two interacting bodies and the distance between them. The force between the masses, say M and m, separated by a distance, d, is proportional to the product of the masses, Mm, divided by the distance squared, d^2. Thus, the force, F, is proportional to [Mm/d^2]. The constant of proportionality is called G, the universal gravitational constant, which is a number that has been measured. The complete equation for the force of gravity between two bodies is:

$$F = G\,[Mm/d^2]$$

We are always aware that if we find ourselves unsupported by a floor or the ground or a step, we will fall. The force of gravity acts everywhere, always, to pull anything down. If our body is not supported, we will fall. That supporting force is what makes our legs tired after prolonged standing.

A convenient way to describe this omnipresence is the idea of a field. The gravitational field assigns to every point in earthly space a force per kilogram of mass, F/m. Using the preceding gravitational force equation, the gravitational field at any point is:

$$F/m = G\,[M/d^2]$$

This concept of a gravitational field enables us to talk about the effect of gravity in space without reference to the falling mass, m. The term *field* will be used often in our discussion of electricity and magnetism: thus, electric fields and magnetic fields.

Newton also produced three laws of motion[37] that enable us to predict the motion of any body subject to any applied force, gravitational or

---

[37] See Appendix at the end of this Chapter.

otherwise. These laws of motion apply universally to the motion of any mass subject to any force.

Finally, Newton developed the mathematics necessary to execute the solutions to his laws, the calculus. The German mathematician and philosopher Gottfried Wilhelm von Leibniz also, independently, developed the calculus.

Newton published his gravitational law and his other findings in his *Principia Mathematica* in 1687. He actually had worked out much of his theory several years before publication.

Newton, the first of the great theoretical scientists, is regarded by many as the greatest. His body of work was complete and profound. He explained much that the astronomers of his time could see, and he rendered motion open to straightforward analysis. He not only provided the substance of a physical law but also developed the mathematics needed to express it.

Newton's work was a marvel of descriptive economy—a few sentences (or equations) give everything that one needs to know about mechanics. Newton's work gave experimental science a premier position in intellectual and, indeed, practical life.

The sensational output in the sciences in the sixteenth and seventeenth centuries exerted enormous influence. Science was to have a deep impact on history. Newton's science, coupled with the explosion in literature and the arts, ended the more-than-a-millennium-old leadership of monastic scholarship.

Pascal's France

Blaise Pascal was ten when Galileo was censured in 1633. His family was well connected with the court during the time of expanding monarchical power. He lived under King Louis XIII and his notorious first minister, Richelieu, whose administration was focused completely on the expansion and solidification of the king's power at the expense of the lesser nobles. He was continually expanding the taxing power of the monarch to the detriment of the lesser nobles as well as the people. Richelieu died in 1642 and Louis XIII died in 1643.

Louis XIV, at age five, ascended to the throne. The new first minister, Cardinal Jules Mazarin, had to focus initially on managing the regency with Queen Anne and also neutralizing the lesser nobles who bristled under the absolute power of the king. Mazarin was able to quell a rebellion against the monarchy, the Fronde, (1648–53). Louis XIV, the *Sun King,* ruled for seventy-two years in all, eighteen under the regency of his mother until 1661, and thereafter fifty-four years personally, until 1715. In his long reign, he defined the absolute monarch, *"L'etat, c'est moi."*

Pascal's father, Etienne, owned a noble title, *noblesse de robe,* which made Blaise a member of the privileged class. His father was a member of the French Academy, a group representing the best scientific minds in France. The Academy was set up by the monarchy under a Franciscan Friar, Marin Mersenne, to advise the court on the state of scientific progress in Europe.[38]

Blaise had a unique window into European culture, as well as science. In his work on the vacuum, he had conceived and executed his experiments while also conducting a dialogue on the reality of a vacuum with Fr. Noel, S.J.[39] who objected to the existence of the vacuum on philosophical grounds.

In this very civil exchange, Pascal argued strongly for the priority of experimental means in matters of physical reality. He respects the legitimate claims of philosophy but rejects its intrusion into matters that can be studied experimentally. The dialogue shows Pascal's polemic skill and also suggests that, despite the misfortunes of Galileo, his contemporaries were quite able to defend the experimental method.

---

[38] Mackenzie, Charles S. *Blaise Pascal, Apologist to Skeptics,* University Press of America, Inc., 2008, p. 34

[39] Ibid. pp. 43-44

Descartes did, however, withhold publication of his *Treatise on the Universe* in 1633, at the time of Galileo's abjuration.[40]

Between Two Lives

Blaise Pascal and Alexis Carrel might be said to have much in common. Both were French scientists of significant accomplishment. Both had been raised as Catholic Christians in *bourgeois* families, and each had lost a parent in his youth. Pascal was the middle child and had two sisters. His father directed his education in detail. He was essentially home-schooled. Pascal never married and was very sickly throughout his thirty-nine-year life.

Carrel's father, who came from a line of textile merchants, died when Alexis was five. His mother enrolled him, the eldest of three, in Jesuit schools. He attended medical school at the University at Lyons. Carrel married Anne de la Motte de Meyrie, a widow with one child, when he was forty. They had no children together. He seemed to enjoy good health and died at seventy-one.

Pascal and Carrel, however, though both were French, lived in very different cultures. The science of Pascal was able to accommodate a philosophical sense and a religious sense. Carrel, however, struggled to articulate any source of knowledge other than science. What separated these scientists was two centuries of French history, in which the French Revolution was the major event.

The cataclysmic changes in religious faith, intellectual life, economics, and politics that characterized modern Europe were the diverse engines by which the new "order" emerged. It is impossible to fashion a chronological story coherently. Events took place contemporaneously and framed each other in a chaotic or impulsive manner. The imperative was to stop

---

[40] Scruton, Roger, *A Short History of Modern Philosophy,* Routledge Classics, 2nd Ed., London and New York, p. 29

the profligacy of an isolated and impermeable monarchy. The object was achieved but at a terrible cost.

In the 211 years between their lives, French culture fragmented in recurring cycles of frenzy and miasma, chasing elusive ideals and then nursing fresh wounds. The soul of France was rent as violent events and fear played on and deepened factional hatreds. Patriotic fervor was often roused by opportunists to spew their particular pathology.

Modern France and her neighbors still seek a European culture based on rejection of the old. The isms of the twentieth century speak definitively of their failure. The old is forgotten but not gone. The revolutionary goals of *liberté, equalité, fraternité* have meaning only in the human heart. Political constructs can neither provide them nor guarantee them.

Appendix: Newton's Three Laws of Motion[41]

1. A body at rest will remain at rest. A body in motion will remain in motion in a straight line, with undiminished velocity, unless influenced by an unbalanced force.

2. A mass under the influence of an unbalanced force will accelerate, i.e., change its velocity at a constant rate in the direction of the unbalanced force. The magnitude of the acceleration is proportional to the ratio of the force to the mass, i.e., $a = F/m$.

3. A body of mass, m, under the influence of a force, F, will exert a force of the same magnitude against the applied force. This is seen commonly in the recoil of a firearm. The third law is commonly stated: "For every action there is an equal and opposite reaction."

---

[41] For a thorough discussion of the application of Newton's laws of motion, see Sheldon L. Glashow, *From Alchemy to Quarks,* Brooks/Cole Publishing, 1994 pp. 92-94

Chapter 3

The Loss of Christian Unity

THE RENAISSANCE

New economics in the fourteenth century made it possible for many to participate in the intellectual life of Europe. Wealth from expanding trade gave cities, notably Venice, worldwide influence and drew tradesmen and professionals to urban centers, where universities arose. The bourgeoisie, from artists to artisans, had time and resources to devote to education and the practical and fine arts. This period of renewal, known as the Renaissance, started in Italy. Italy, which had been the repository for most of the remnants of classical history, benefited earlier from the economic boom. Wealthy Italians were disposed to invest heavily in the arts.

Italians dominated Renaissance art with painters and/or sculptors such as Giotto and Raphael, da Vinci, Michelangelo, Ghiberti, and Donatello. Spain produced El Greco; the Low Countries, Rembrandt and Rubens; and Germany, Dürer and Holbein. These artists produced prolifically and made major advances in technique. Architecture also flourished as increased wealth stimulated building throughout Europe.

Writers revisited the Greek and Roman classics, which were, if not "newly discovered," at least broadly available to lay scholars. Authors of the "new learning" not only wrote in the traditional Latin but also in the vernacular languages, Italian, English, Spanish, and French. These authors produced classic works in literature, in social commentary, and in the arts. The invention of the printing press in the mid-fifteenth century created widespread demand for both the ancient classics and for the works of Renaissance authors.

The Divine Comedy of Dante Alighieri (1265–1321), written in Italian at the beginning of the fourteenth century, we may say, set off the Renaissance. Dante, accompanied by his beloved Beatrice, makes a tour of the three venues of the hereafter, *Paradiso*, *Purgatorio*, and the *Inferno*. The tour is an exposition of the temper of the times by an author who was both a serious believer as well as a political polemicist, exiled from his native Florence. On the boundary of human history and eternity, we visit well-known individuals, each in his eternal home. The interest lies in Dante's view of where each figure belongs and why.

Geoffrey Chaucer (1342–1400) wrote his *The Canterbury Tales* in English later in the century. It is an entertaining, sometimes bawdy, collection of stories, recounted poetically by travelers of widely disparate sensibilities on a pilgrimage to a shrine at Canterbury. This poetic masterpiece still compels attention and provides a marvelous example of the art form. Such contributions from lay artists, their mastery of language, their insight into human foibles, and the beauty of their art offer compelling evidence of their genius.

Two very close friends, Sir Thomas More (1478–1535) in England and Desiderius Erasmus (1466–1535) in Holland, wrote extensive commentaries on intellectual life in Latin. They were aware of and encouraged growth in religious fervor and holiness in laypeople. While they strongly embraced inherited medieval traditions, they were devoted to the "new learning" and lived lives committed to prayer and the sacraments. They lamented the inertia and smug dogmatism of the scholastic schoolmen and theologians and spent great effort in the study of the classics and the new scholarship in literature and the sciences. These authors were committed Christians whose goal was sanctification.

Miguel de Cervantes Saavedra (1547–1616), the major Renaissance author in Spanish, wrote what is believed by many to be the first and the greatest novel, *Don Quixote*. The work is an imaginative and powerful satire that plays on the complex variety of social strata and castes of seventeenth-century Spain, heavily influenced by diverse religious sensibilities.

Don Quixote is a manic figure who sets out to practice the chivalrous code of the knight-errant. The fact that feudalism and chivalry exist only in fantasy sets the stage for Quixote's folly, which makes the hapless knight a laughingstock in seventeenth-century Spain. Cervantes, who was influenced by Erasmus's individual piety, uses several very bold literary devices, such as effectively switching authors within the text, to scramble the motivations and behavior of the characters. The musical comedy *The Man of La Mancha* is an adaptation of *Don Quixote*.

The greatest playwright in the English language, William Shakespeare (1564–1616), was a contemporary of Cervantes. Shakespeare's poetic dramas and sonnets were written in the time of the Tudor Reformation in England. Shakespeare, a man of Stratford-on-Avon, did most of his work in the theater in London. He was living when Elizabeth I was firming up the installation of the Anglican Church in England. Shakespeare probably set some of his plays in foreign or fictional countries in order to maintain artistic freedom and a low profile. His modest education has led some to wonder if he authored the works attributed to him, but it seems far-fetched that he had or was a ghost.

Moliere (née Jean-Baptiste Poquelin) (1622–1673), a notable French dramatist, was a contemporary of Pascal. His repertoire includes many farcical comedies skewering social customs and personal excesses in France.

THE SPIRIT OF THE RENAISSANCE

In the mid-fifteenth century the printing press gave impetus to this immense literary effort. The number and quality of new authors bred wide readership.

The breadth of interest sprang from an intellectually disposed bourgeois culture. Dante, Chaucer, More, Erasmus, Cervantes, Shakespeare, and Molière created wondrous literature. Science also emerged from the Renaissance as a parade led by Isaac Newton. The development of a new intellectual class fostered a spirit of openness and discovery.

Dante's *Inferno* is a stark insight into the scandalous admixture of the divine and the profane in fourteenth-century Italy. The poet, at midlife, is reflecting upon his own state of soul. On this tour of hell, Dante, guided by the Roman poet Virgil, witnesses the fate of historic figures known for ambition and corruption. He also dialogues with his own deceased contemporaries, whose venality or duplicity he knows firsthand. Within the ranks of the damned are Christians, from popes and princes to intellectuals and hirelings. In each case, the punishment is suited to the offense. The man with temporal and spiritual duties encounters intractable challenges in attempting to serve two masters. Dante suggests it would be difficult to imagine that the counsels of Christ could be practiced in the environment of many medieval churchmen.

Erasmus and More, as lay Catholics, in particular strove to form a devout Catholic laity. Thomas More's *Utopia* portrays an idealized society, patterned on Christian values. His martyrdom continues to witness to our postmodern world the courage needed to maintain one's faith under extreme pressure.[42]

Don Quixote's hilarity is drawn from Cervantes associating his Spain with the era of the knights errant. Medieval chivalry in Spain's post-Columbus colonial empire is nonsensical. *Don Quixote*, as a laughingstock, is analogous to the "divine right" feudal monarchy that is the government of Spain. Quixote's intense practice of the traditions of knighthood and courtly love is funny, because he is out of touch just like the courtly king and queen.

---

[42] Robert Bolt, *A Man for All Seasons*, Modern Classics edition, Bloomsbury Methuen Drama Publishing, London, 1996.

Shakespeare's plays do not directly address the parochial incongruities of England, for he wrote in a time of horrific persecution. Yet his work is timelessly relevant, for he fixes on the enduring human condition. His drama and poetry set a standard for English literature that endures. From the historical perspective Shakespeare represents a transformation, which can be thought a perpetual standard.

Shakespeare could not craft his plays on religious issues, but given the cultural strictures of Elizabethan England, some speculate that such a genius would leave implicit commentary on persecution in his plays. Many are set in foreign countries, where English strictures do not apply. Some scholars today are inferring that Shakespeare's personal religious commitment is evident in particular characters in several of his plays.[43]

The Reformation

When, in 1517, Martin Luther posted his ninety-five theses on a church door in Wittenberg, Germany, he most likely neither intended nor foresaw the Reformation that he initiated. But within the Church, pastoral ministry was often leaderless. Intellectual and disciplinary atrophy among the clergy and a lack of pastoral focus by the hierarchy were routine.

Christ proclaimed he had no place to rest his head. Yet many pastors with benefices abandoned their parishes. Bishops were isolated from their priests. The training of curates was marginal. Class consciousness soured the Church. Luther simply nudged the first card and the unstable system toppled.

Religious Wars

Luther's challenge triggered diverse responses, generally in northern Europe, where political and economic interests were quick to make use of religious

---

[43] See Joseph Pearce, *The Quest for Shakespeare*, Ignatius Press, San Francisco, 2008.

factionalism. Major political changes such as the centralization of monarchical power were conflated with religious animosity and mistrust. There were civil wars in France and Germany and wars between England and Spain and then between Spain and Holland. Religious wars persisted through most of the sixteenth century.

The Saint Bartholomew Massacre

Interreligious animosity occurred later in France than in the rest of Europe. When enmity arose in the 1560s, Protestant Huguenots, primarily the merchant class, and the Catholic monarchists inflicted carnage on each other sporadically over about forty years. The most notorious eruption, the St. Bartholomew Massacre, occurred in August 1572.

Queen Catherine de Medici, mother of King Charles IX, had arranged a state marriage between her Catholic daughter, Marguerite, and Henry, the Protestant king of Navarre, who was heir presumptive to the throne. The queen's intention was to promote tolerance and eventual unity between Catholics and Protestant Huguenots. She hoped that love would conquer all.

The hatred between Catholics and Huguenots festered anew as the two groups assembled in Paris for the marriage in August 1572. As the hour approached, antipathy intensified, becoming so strong that the sponsors despaired of controlling it. The fury of Catholics, unchecked by the authorities, was unleashed on the assembled Huguenots. An estimated five thousand were killed.

The duplicity and bloodlust of the St. Bartholomew Massacre poisoned French society. When, in 1589, Henry of Navarre became King Henry IV, as a gesture of reconciliation he became a Catholic. In 1598 he was able to interrupt most of the futile hostilities by negotiating terms of toleration for the Huguenot minority in the Edict of Nantes. But the memory of past duplicities and killing sprees was very strong. The French monarchy could not erase the pervasive distrust, but it sought to prevent a relapse at all costs.

This history was a major threat to stability. Determination to avoid religious strife became an imperative of French politics.

Pascal was born in 1623, about midway in the Thirty Years War, a multinational, sporadic conflict fought mostly in Germany, initially between Catholic and Protestant countries. In fact, as the war progressed, nations switched sides for balance-of-power reasons. This war, which is perceived as a "religious war," evolved into lineups that had co-religionists on opposite sides. Religion as religion may not have been the cause of any given war, but religion was clearly used by princes to enflame hatreds. The Treaty of Westphalia ended the conflict, the last of the major wars of religion, in 1648 when Blaise Pascal was twenty-six.

Mère Angélique and Port Royal [44]

In 1602 an abbess was assigned to a convent (Port Royal des Champs) of twelve nuns twenty miles south of Paris. The event was, at the time, routine, but it was both bizarre by any sensibility and pivotal in the history of France. The event was bizarre because it shows the shenanigans possible for the privileged class in seventeenth-century France. The central figure is Jacqueline Marie Arnauld, the new abbess. At the time of her appointment she was ten years old—almost eleven. This corruption exemplifies the secular impositions on church matters that could be arranged.

Antoine Arnauld and his wife, Catherine, had twenty children, ten of whom survived. A noted attorney of noble rank, Antoine could not provide a dowry for all his girls, so he planned to marry off the eldest, Catherine, and the others would be nuns. Jacqueline Marie, Antoine's second daughter, was the first to enter the religious life, as Mère Angélique.

At her appointment, the community consisted of twelve adult women, whose discipline was informal, or perhaps *flexible* is a better term. Antoine

---

[44] O'Connell, *Blaise Pascal: Reasons of the Heart*, pp. 57-64.

procured the benefice for Jacqueline Marie, getting the necessary permissions by telling some judicious fibs. As a preteen or "tween," as they say today, she was installed as the Abbess Mère Angélique.

Mère Angélique initially went with the flow and kept to herself, but in her mid-teens, during Lent, she did some spiritual reading. She undertook some penitential practices, which fostered a sense of responsibility within her. She was moved by the advice of two itinerant preachers to consider some reform. Consultation with her Cistercian superiors convinced her that they had no interest in reform. But she also got the idea that they were not going to interfere either.

So she set out, by example and incrementally, to influence the culture of her convent. As her more mature sisters saw the transformation in their young abbess, they were moved to follow her lead. As she grew in their regard, she introduced one after another of the traditional elements of work and prayer into the community. With patient resolve she was able to convert a monastery, which had abandoned the Benedictine rule, into a model of the original Cistercian discipline.

Along the way, her father and mother, who had free access when visiting, were received in a reception area near the gate. Mère Angélique had instituted cloister. Her father knew she was within her rights and observed the rule. Port Royal des Champs became noted for its devotion and discipline.

In 1625 Mère Angélique moved the convent to Paris. The original location, Port Royal des Champs, was inadequate for the growing population of the convent. The new location, Port Royal de Paris, was established. The convent was also released from Cistercian oversight, becoming responsible to the Archbishop of Paris. The Arnauld family by this time was heavily represented at Port Royal. Mère Angélique's widowed mother, Catherine

(Arnauld) and her widowed older sister, Catherine LeMaître, as well as four other sisters and six nieces, were nuns at Port Royal.[45]

Mère Angélique had been introduced to a priest, Abbé Saint-Cyran, at Port Royal des Champs by her older brother, Robert d'Andilly. She had reservations about Saint-Cyran, who had preached there on a few occasions, because he was the absentee bearer of a benefice from another monastery. In 1635 she had occasion to appoint a new spiritual director for the community. Her experience with spiritual directors had been mixed, and she was not one to follow first impressions, often finding that appearances were deceiving. But the years of Saint-Cyran's association with the various members of the family inside and outside the convent led her to select him as spiritual director.

Saint-Cyran was very effective in winning the admiration of the Port Royal community. However, his influence with the elites of Paris led to severe political consequences in 1638, when he was arrested on Cardinal Richelieu's order and imprisoned in Vincennes. The first minister had found Saint-Cyran's pastoral influence within the royal household was interfering with matters of state. This charismatic priest was apparently disturbing consciences. Richelieu recognized no boundaries when matters of state were involved.

Following Saint-Cyran's arrest, Catherine LeMaître's son and Mère Angélique's nephew, Antoine LeMâtre, another brilliant young attorney, raised Richelieu's ire by publicly defending Saint-Cyran, as did many others, such as Francis de Sales. Antoine, thereupon, abandoned law for a life of piety. His brothers, Isaac and Simon, followed him in short order. They were

---

[45] Robert Arnauld d'Andilly was the eldest son among the ten survivors of the senior Antoine and Catherine Marion Arnaud's twenty children. Their first daughter, Catherine LeMaître's, was widowed in 1619. Jacqueline Marie, Mère Angélique, was the second daughter. There were then three girls, the fourth, fifth, and sixth children. The last four—seventh, eighth, ninth, and tenth—were boys. The tenth and youngest being Antoine, the Great Arnauld. He was to be a combative and embattled theologian and author.

eventually joined by their widowed uncle, Robert d'Andilly, all taking refuge in Port Royal des Champs. These men, all Arnaulds,[46] formed the core of a lay religious community, the *Solitaries*, who refurbished the old monastery and undertook a variety of apostolic tasks, most notably education.

JANSENISM[47]

Saint-Cyran is a pivotal character in that he introduces Jansenism into our story—it is a major element historically. Saint-Cyran had been born Jean Ambroise du Vergier. Saint-Cyran is a name given with a benefice after he had been ordained in 1618. Recall the Pascals' pastor in Rouville, Curé Jean Guillebert. He was schooled in Jansenism by Duvergier.

Although Duvergier and Cornelius Jansen (also known as Jansenius) both attended the university at Louvain in the Spanish Netherlands at the same time, they did not interact until they had met in Paris taking advanced study in 1609. From 1611 until 1616 Duvergier and Jansen collaborated in the study of St. Augustine's soteriology, the theology of salvation. These studies are relevant to the sacrament of reconciliation, confession, or penance.

When, in 1646, the Pascal family decided to live their faith more deeply, they reflected a trend that was sweeping much of Catholic Europe. The Counter-Reformation had been formulated at the Council of Trent (1545–1563). The Counter-Reformation spurred Catholic laity to a deeper devotion to Christ and a renewed practice of charity. New religious orders were formed to serve the poor, to teach, and to evangelize in Europe and around the world. Bishops were now required to serve, govern, and reside in their dioceses. Seminaries were reformed to provide clergy with a better

---

[46] Arnaulds at Port Royal: Mère Angélique's mother, Catherine Marion Arnauld; four sisters, Catherine LeMaître's and three younger girls; six other nieces. In addition, the three of her four younger brothers; her nephew, Antoine LeMattre, the former lawyer, her older widowed brother, Robert Arnauld d'Andilly, all took refuge as *Solitaires* at Port Royal des Champs.

[47] O'Connell, pp. 34–42.

preparation in philosophy and theology. Pastors were required to personally guide their parishioners in worship: prayer and the sacraments. The Catholic laity was inspired to a deeper practice of charity and the devout life.

There is an account in St. Matthew's gospel (Matt. 10:17–22) where Christ is asked by a rich young man, "What must I do to gain eternal life?" Christ's answer is to keep the commandments. The young man responds, "Ever since I was young, I have obeyed the commandments." Christ, with love, responds, "You need only one thing. Go and sell what you have and give the money to the poor . . . then come and follow me." The young man turned away sadly, for he was very rich.

This account sharply distinguishes between a minimal standard of Christian piety and the higher ideal of union with God. Avoiding sin is necessary for salvation, as a minimum. If a person, progressing against the world, the flesh, and the devil, wishes to know, love, and serve God more deeply, then he must remove attachments to earthly things and do the will of God as Christ reveals it to him in the circumstances of his life.

Everyone must, as a minimum, obey the commandments. Such a soul will avoid damnation. But if one wants a deeper union with Christ, then one must give up legitimate human interests, forgoing one's own will and adopting the will of God, i.e., to do the work of God.

The first level is necessary. The second level is a counsel, not necessary but, as Christ called it, "the better part." Christian piety, if begun, holds out to everyone this "better part," deeper union with God.

As fallen human beings, we all sin. The sacrament of reconciliation, confession, or penance is the sacrament by which the Catholic Church administers Christ's mercy through His forgiveness of sin. As with Augustine, so we all can receive absolution of our sins if we confess all our sins, are truly sorry, and firmly resolve to reform our lives.

An especially fervent soul, aspiring to conform his will to Christ's, is aided by the sacrament in searching for dispositions of heart that keep him from a virtuous life. His motivation is a greater love for God. Such devotion is commended to all, but forgiveness is available to any penitent who has sincere sorrow for whatever reason—shame or fear of hell, for example.

In very simple terms, Jansenism seemed to require all, even the morally weak, to have the higher level of holiness to be forgiven. In confession, this standard meant sorrow for the love of God was required not for baser motives like shame or fear. Such a broad extension of the ideal puts an unnecessary burden on the weaker penitent. Anyone can fall, and shame and/or fear will serve for forgiveness. Christ calls each and every person to repentance at whatever level of sorrow he has, as long as he confesses all his sins and sincerely wishes to change his behavior.

Jansenist theology also describes specific characterizations of the grace God allots to the soul, how grace acts within the soul, and how free the soul is. The Jansenist posture could be conflated with the Calvinist idea that God predestines souls to heaven or hell. The Counter-Reformation Church was very sensitive to interpretations that could be misconstrued as supporting predestination.

The Suppression of Jansenism[48]

When, in 1602, Antoine Arnauld purchased the Port Royal benefice for his daughter, Jacqueline Marie, he could not have imagined what would emerge from his initiative. His youngest son, and Mère Angélique's brother, Antoine, a gifted lawyer, emerged as a brilliant and combative theologian. He was ordained a priest and received a doctorate in theology from the Sorbonne in 1641. Known as the Great Arnauld, he became the lightning rod for the political and theological firestorm that was to come down on Port Royal.

Three years earlier, in 1638, Antoine, who knew the Rouville Curé Jean Guillebert from the Sorbonne, had communicated his esteem to the imprisoned Abbé de Saint-Cyran and sought his spiritual direction. At the time,

---

[48] O'Connell, chapter 7.

Saint-Cyran was imprisoned by Richelieu. Saint-Cyran had been affronted by information he had of laxity among Jesuit confessors. He told Antoine that he understood the Jesuits were encouraging people living morally superficial lives to receive the Eucharist without what he considered sufficient sensitivity to the gravity of their sins. In prison Saint-Cyran could not act on this information, but he sought to interest the young theologian in accepting that assignment. Antoine agreed and was poised to champion Saint-Cyran's brand of penitential contrition in administration of the sacrament of reconciliation.

Jansen died in 1638, and his book of theology, commonly titled *Augustinus*, was published posthumously in 1640. Saint-Cyran had been, as Duvergier, Jansen's theological collaborator. Jansen's book presented his theology of salvation, formed by applying the new "positive theology" to the writings of Augustine. As scholarship, it stimulated detailed discussion among theologians.

Jansen's scholarship may have been a reasonable interpretation of the text standing alone, but its use as a means to infer a standard for the forgiveness of sin in confession was stretching a Christian counsel into a standard for all. Saint-Cyran had pride of ownership in *Augustinus*. but as a teacher, preacher, and spiritual guide, it was not Christ's charge to make the "better part" the only part.

When Richelieu died in 1642, his successor as first minister, Jules Cardinal Mazarin, would take some time to fill the vacuum created by Richelieu's departure. Shortly thereafter, in 1643 King Louis XIII died, leaving Queen Anne, mother of the five-year-old king, Louis XIV, and Cardinal Mazarin to set up and manage the regency. Mazarin was not ready to react to religious controversy, but his Gallican sense recognized its potential for political mischief. Saint-Cyran died shortly after his release from prison in 1643.

In that same year, the Great Arnauld made good on his promise to Saint-Cyran. He threw down the gauntlet to the Jesuits in the Sorbonne by publishing *De la Fréquenie Communion* in reaction to their alleged softness on *honnête homme* penitents. The publication exhibited the power of Arnauld's facile intellect and persuasive rhetoric.

A pamphleteering war over Jansenism festered for the rest of the 1640s. Support for the Jansenist position was not extensive but limited to a number of influential and committed partisans. There was a core of the Saint-Cyran followers in the court and among the bourgeois and the nobility.

Yet there was strong opposition from some churchmen in pastoral work. For example, Vincent de Paul, a national Counter-Reformation icon for his heroic work for the poor and the epitome of charity, was strongly opposed to Saint-Cyran's ideas. The academic theologians and the political and religious hierarchies were generally not supportive.

Cardinal Mazarin, already distracted by administering the regency for Anne of Austria, had to deal with remnants of the French nobility concerned with the growth of monarchical power. The *Noblesse d'épée*, sensing a time of weakness, were harassing the crown. For about five years starting in 1648, they precipitated political instability by a series of uncoordinated attacks, called the Fronde—some military, some clandestine against the monarchy. Mazarin was up for the challenge, but it distracted him from focusing on the Jansenist dispute for some time.

In 1649 the censor of the faculty of theology at the Sorbonne issued a recommendation that five propositions from *Augustinus* and two propositions from *De la Fréquente Communion* be designated heretical. Antoine Arnauld, a sitting member of the faculty, was facing repudiation and removal. In short order, a review by the Assembly of French Clergy recommended that the two propositions in *De la Fréquente Communion* should be excluded from consideration, thus removing the propositions due to Arnauld.

The stakes were raised when several bishops appealed to Rome in 1650 to consider the propositions. The Pope appointed a study commission, which within a year upheld the charge that Jansen's five propositions from *Augustinus* were indeed heretical.

The process in the Sorbonne was prolonged when Antoine Arnauld, while agreeing with the Pope's decision, contended that the statements attributed to Jansen did not, in fact, appear in *Augustinus*. In other words, the Pope was right in finding the propositions heretical, but they did not come from *Augustinus*. This distinction was defended vigorously by Arnauld and his supporters but ultimately was rejected by the Assembly of French Clergy in early 1654. Two appeals to the faculty at the Sorbonne were rejected, and in January 1656 Arnauld realized further argument was useless.

The fact that Arnauld's father, as an attorney in legal proceedings in 1590, had facilitated the expulsion of the Jesuits from the Sorbonne under Henry IV is said to have inspired payback for the son. But the Great Arnauld had done a very good job of poisoning the well on his own. This animus would endure to become a blot on French history and would ultimately inflict serious harm on the country and the Church at the time of the French Revolution.

The Provincial Letters

The Pascal family's decision to accept the brothers Deschamps' invitation to a deeper religious practice was motivated solely by their desire to do God's will more faithfully. The maneuverings of the Great Arnauld against the Jesuits on behalf of Saint-Cyran were not on Pascal's radar. The interplay of theologians and the animus that was fostered over Saint-Cyran's ministry was an intrusion to them. But Blaise Pascal and his sisters, Gilberte and Jacqueline, had strong ties to the Port Royal community and the friendships they found there.

When Antoine Arnauld was facing removal from the theology faculty of the Sorbonne, the theological issue was a focus of grudge and payback on both sides. Pascal was not a theologian, but he was an exceptionally effective polemicist. He wrote his *Provincial Letters* as a defense of his aggrieved friend and the Port Royal community. The *Letters* were a series of essays written over fifteen months (January 23, 1656 to March 24, 1657) and were the culmination of a long, drawn-out public dispute on the issue of Saint-Cyran. Pascal's rhetorical skill made the *Provincial Letters* a masterpiece of French prose. Independent of the merits of the Jesuit position, Pascal's studied characterization of their behavior was withering.

The *Provincial Letters*, published anonymously, were written as letters by a man from the provinces to a friend back home. Coyly affecting the common sense of the heartland, he described with feigned confusion what was happening at the Sorbonne. He portrayed the Jesuit protagonists by recounting conversations he had with them and others, seeking to understand the proceedings. His parodies of Jesuit theologians were crafted to imply all sorts of opportunistic and shady dealings in the mind of the man back home. Pascal's polemical skill heaped scorn, tarnishing the perception of Jesuits. His irony and sarcasm made the word *Jesuitical* synonymous with "cunning or deceitful."

The "Great Arnauld" did not prevail. Pascal's letters, after the first three, have a tone of resignation, but they were aimed at the Jesuits and their methods, not the question that had been at issue. Pascal's rhetorical skill was turned in full fury on the Jesuits, and the sting was remembered.

THE FRENCH REVOLUTION

THE STATE OF THE NATION

The French Revolution cannot be neatly packaged or rationally reconstructed. The mood in Paris and in the provinces was outrage. France overwhelmingly and diversely was demanding an end to the status quo. The revolution began as an attempt to solve an economic problem, bankruptcy. But the indolent King Louis XVI provided no executive leadership to define the problem or offer direction. The process was bedeviled by untimely events that removed options and changed the landscape.

The monarchy, leading up to the revolution in eighteenth-century France, had done nothing to stifle the seeds of revolt. From 1715 until 1774 Louis XV—an indolent, self-absorbed wastrel—carried on a profligate life fully indulging his whims. The excessive spending went on, creating a larger privileged class, further isolating the monarchy from social, economic, and political reality.

Bankruptcy and confiscatory taxes were the price of pompous trappings such as Versailles and wars, never-ending wars. The intractable, unjust, and outmoded tax structure had to fall.

Louis XVI, who inherited the throne in 1774, was not a profligate spender, but he had absolutely no capacity for leadership. He was inept, insecure—unable to use his power and French patriotism to motivate his people. He muddled along, creating a dangerous vacuum, opened to diverse and unpredictable opportunists.

In 1789 he was forced to recognize that France was bankrupt and that substantial change was unavoidable. He was not prepared to propose a solution, nor offer a set of choices. Instead he punted, seeking a consensus by convening a meeting of the Estates General.

The Estates General represented the three "Estates" of French society—the church, the nobility, and the common folk. But it was advisory and had no legislative power, nor was it an appropriate vehicle for executive decisions.

The First Estate was the church. The church was not only the titled bishops and abbots who functioned as clerical nobility but also the vast number of poor curates who did the lion's share of ministry to the faithful.

The Second Estate was the hereditary nobles, *Noblesse d'Epée*, and the bourgeoisie, *Noblesse de Robe*, who bought their titles.

The Third Estate included all who did not have clerical status or a title—the peasants, artisans, shopkeepers, artists, professors, businessmen, etc.

As constituted, the Estates General was also a stacked deck. The commoners, the Third Estate, were window dressing. Assuming the First and Second Estates each had one hundred elected delegates, then the Third Estate would have two hundred delegates. But votes were cast within each Estate separately. Proposals required a majority vote of the Estates—each Estate having one vote. So the First and Second Estates together could prevail over the people's Third Estate. Theoretically 102 individual votes—51 each in the First and Second Estates—could outweigh 298 votes in all three Estates.

The Course of the Revolution, 1789–1795

The National Assembly

The elected representatives of the Third Estate were determined to remove the voting sham. As they gathered in Paris, the Third Estate representatives, with virtual unanimity, demanded a change in the voting system. When the delegates arrived in May 1789, the Third Estate refused to participate unless the biased voting system was revised. Their input must have some teeth, if substantial reform was to be enacted. A significant minority in the other Estates supported the demands of the Third Estate.

When the king failed to respond in a timely way, the Third Estate jumped into the void, declaring itself a national assembly and inviting representatives of the other Estates to join them in forming the assembly. Deliberations would result in resolutions determined by a majority vote of the whole assembly. The assembly would then create a constitution for a limited monarchy.

In late June the dawdling king agreed to the national assembly and its voting procedure. His passivity in not engaging the representatives in specific preparatory understandings again created an executive void. With the voting system changed, some kind of reform was assured. But what kind? Direction from the king to the national assembly was unforthcoming. His maneuvering at a distance enabled the most organized and willful to fill the void.

Most of the representatives were motivated by a sincere desire to bring their country back to sustainability—not to dismantle it. However, the unanimity seen in the matter of deliberative procedures in the national assembly virtually disappeared in considering how reform would be executed. When it came to the details, the pain of old sores was the strongest motivator. A small but organized minority stood ready, in time, to direct a "popular revolt."

A king with courage and insight would have checked the excesses within the national assembly. Louis XVI squandered his power to rally French patriotism by defining the sacrifices he deemed reasonable.

The people of Paris, who had been organized to conduct the election of delegates to the Estates General, redirected their interests to support their representatives in the deliberations of the national assembly. There was similar militant interest throughout the country and local authorities had to deal with generalized restlessness among the people. But Paris was the volatile arena within which the collapse would proceed.

Anxious Parisians assumed a crackdown was near. Some did not intend to wait passively. They hoped their vigilance might detect a marshalling of force within the city. They began to monitor activity around local police and military garrisons.

On July 14, 1789, a crowd approached the bastille, a prison and military post in the city, to see what was up. Their representatives were received and invited inside to have a look. They found no massing of force, but they tarried a bit. When they did not return in what the crowd thought was a reasonable time, some daring souls scaled a wall and opened a gate. The crowd poured in. Their intrusion caught the authorities by surprise, and a battle ensued. By the end of the day, with support of armed forces loyal to the group, the bastille fell. The citizens at the bastille, in fear and distrust, had done what they thought they should do there and then. Their act, as a symbol of resistance, galvanized a revolutionary zeal throughout France. The tinder was dry when the match was struck.

The Assault on the Church

Now revolutionary fervor was fired by a battle won, a minor encounter but an enduring symbol. More importantly, the street people had become a political force. The national assembly now had to deliberate under intimidation by the Paris mob, which could be called out as needed.

Now the radical element in the assembly could notch up its effort to remake France to its own liking. To pay off France's debt, the church expected the assembly would confiscate a great deal of the land that had accrued to the church under feudal-like relationships. The landed nobility was equally subject to such confiscations. A just and equitable system of confiscation was the first order of business.

However, the bishops did not know that now "the admitted necessity of these ecclesiastical changes should be seized upon as a pretext and a handle by men whose purposes were simply destructive; who designed not only to break down the external barriers of an enormously wealthy church establishment, but also to level with the ground the inner citadel of Catholicism, and even Christianity itself."[49] This "inner citadel" of Catholicism is the ministry of the Church to its people with the authority of Christ Himself.

To establish an important distinction, the lowercase *church* refers to a proper political identification, and the uppercase *Church* refers to the Mystical Body of Christ.

On November 2, 1789, under the leadership of Gabriel Riqueti Mirabeau, who, after delaying the vote for two days to firm up support, oversaw assembly passage of a resolution that claimed all church lands for the nation. This total confiscation set a tone of punishment rather than sacrifice. Arbitrary power set the stage for animus toward Catholicism that would end in carnage.

The vote on this nationalization of church property was conducted in the presence of "an armed band of lawless miscreants."[50] The assembly was now captive to a radical agenda. A tally of the votes—ayes, nays, and

---

[49] Jervis, W. Henley, *The Gallican Church and the Revolution*, Kegan Paul, Trench and Co. London, 1882, p. 4.

[50] Ibid. p. 39.

abstentions—shows that two hundred of the twelve hundred delegates were absent, some having already left France.

The inability of the church to rally support was not due to active opposition but rather to an endemic identification of the church with the political system. The identity of the church with factional polarizations educed a passive "everyone for himself" attitude among the delegates. The church had created resentments politically, which did not discriminate between the Church in her sacramental ministry and the church in her public and political reputation. Many votes were motivated against a public political-entity church, not the sacramental-ministry Church.

Within the First Estate there was a long-standing isolation between the parochial clergy and their superiors, the bishops, abbots, and other church administrators. The curates were effectively excluded from any chance of higher positions in the church and exclusively "bore the heat of the day" in the actual sacramental ministry of the Church. When issues of church property and monasteries arose, these clerics did not identify with their superiors. The pathology of privilege in French society had created insensitivities and isolations that were now consequential.

One old but persistent animosity that poisoned church interests in the assembly was the suppression of Jansenism. The Great Arnauld was dismissed from the Sorbonne in 1654. Why, then, 135 years later, was the Jansenist wound unhealed? The French ruling system could not limit a strictly theological dispute to a theological solution. Because of the entwining of the church with the power of the state, the capacities of churchmen, Jesuits in this case, were augmented by their identification with the ruling class.

In 1713, fifty-nine years after Arnauld was bounced and the *Provincial Letters* had appeared, a papal document, *Unigenitus*, was issued by Pope Clement XI. It has been characterized as "an act of vengeance directed by the Jesuit, Le Tellier, and his order against those who were bold enough to dispute their [the Jesuits] mischievous ascendancy in church and state."[51] This document led to an active persecution of Jansenists, e.g., denial of sacraments to the dying. A much-loved bishop who refused the required submission was condemned, driven from his diocese, and imprisoned. The efforts by the state to enforce the decree created a fissure that was hard to forget, disaffecting a significant number of people, such as Armand G. Camus, who was in a position to effect legislation as a leader of the National Assembly Ecclesiastical Committee. He too late regretted the effects of his legislation.

The Civil Constitution of the Clergy (July 12, 1790)

The next item on reform was a reorganization of church administrative structure, which put the revolution firmly and irreconcilably beyond reasonable statecraft and guaranteed unrelenting opposition from the Church bishops and clergy across the board. The Civil Constitution of the Clergy interfered with the papal prerogatives of clerical authority and administration of dioceses.[52]

The first provision was that the church dioceses within France would have to coincide with the geographical limits of each of the newly established departments. This was a violation of Church prerogatives and would eliminate fifty bishoprics. But it also required the rearrangement of provinces and parishes, as well as adoption of new clerical titles and the dissolution of local organizations. The law also denied the jurisdiction of any foreign bishop over French subjects, an obvious rejection of the primacy of the bishop of Rome.

A second innovation was the direct election of bishops and parish priests by the people, historic precedence being questionable justification. Again, the legislation denied papal jurisdiction in such local matters. The requirement that French bishops, in fact all bishops, receive "institution," a qualification from the Pope to exercise his office, was abrogated. Similar qualifying documentation of parish priests, through their bishops, was abrogated.

---

[51] Ibid. p. 14 ff.

[52] Ibid. pp. 57–72.

Armand Camus, a leading Jansenist, was influential in the Ecclesiastical Committee, which produced the proposed Civil Constitution of the Clergy. The persistent persecution of the Jansenists by papal edict very likely motivated their support for this attack on papal authority. This move has been attributed to Jansenist resolve "to take summary and signal vengeance for the bull, *Unigenitus*, and all the miseries which had resulted from it for seventy years past."[53]

The Civil Constitution of the Clergy was approved by the assembly on July 12, 1790, and sent to the king for his approval. Louis XVI was in touch with Rome and was, of course, unsure of what to do. The pope, in a secret venue, made it clear that he understood the details of the legislation and enumerated its negative effects on the Church in France and its interference with the legitimate prerogatives of the Church. He left no doubt of the destructive nature of the law. He would wait for a favorable opportunity to speak publicly.

This Civil Constitution of the Clergy was a major overreach by the national assembly, for it hardened opposition from all classes and the mainstream of France. The radical turn of the assembly was evident. A direct interference with the apostolic ministry of the Catholic Church in France was a malignant sign.

The primacy of the pope does not depend on secular approval. The mission of the Church, to sanctify the world, is given to the apostles and their successors and would not be subject to political domination. Such was the outcome of Roman persecution over several hundred years. The Civil Constitution of the Clergy was an abomination to the vast majority of bishops, priests, and devout Catholics. Any effort to impose such strictures was going to be vigorously and unrelentingly opposed.

The king did not have the fortitude to reject the legislation based on its violation of his own conscience and on the clear indications from Rome that

---

[53] Ibid. p. 65.

its primary provisions were attempts to constrict the teaching authority of the Church, which comes from Christ Himself. Such a posture would have rallied royalist and Catholic morale across classes and would have forced reconsideration. In addition, it would have moved the debate out of the assembly and into the broader venue, where mobs and political prejudices are less important. But the docile king approved the measure on August 24, 1790. The immediate effect was to dishearten his Catholic supporters and to embolden the most radical elements of the revolution. It also was a clear indication that the Second Estate royalists were next.

Regicide and the Reign of Terror

The king, given what he was, and given the call of Queen Marie Antoinette for repression, took stock of his situation and decided to leave France, to abandon his people and flee to the Austrian Netherlands on France's northeast border, where his wife's brother, Leopold II, was emperor.[54]

The royal party left in disguise at night and was to link up with the Austrians at the border. The party was captured at Varennes on June 25, 1791. The king was returned to Paris a virtual prisoner. His failed attempt to flee raised the possibility of foreign intervention.

The national assembly came up with a constitution for a limited monarchy in September 1791, which the king duly endorsed. But this king was not going to be their monarch under any circumstances. The assembly, having done its work, ordered the election of a national convention, which would create a constitution for a democratic republic, The assembly was dissolved and a national convention convened in September 1792.

Louis XVI had been deposed by the assembly on August 10, 1792, and was beheaded on January 21, 1793.

The convention produced a constitution for the First French Republic in June of 1793. The convention, however, had to address perceived military

---

[54] Rudé, George, *Revolutionary Europe, 1783–1815*, Harper & Rowe, 1965, p. 125.

emergencies. It remained in power to counter revolutionary threats from neighboring monarchs and to quell any French source of resistance.

Over a period of about two years, the national convention's Committee on Public Safety murdered many French citizens. Anyone associated with the *ancien régime* was subject to elimination. Anyone who was deemed deficient in loyalty to the revolution was at risk. In such an environment, "old scores" were settled and perceived "enemies" of every sort could be dispatched. The emotional ferocity of the time, the Reign of Terror, bred hatreds that poisoned France for many years. France managed to prevail, but its fractured psyche was bereft of reason for some time.

Empires, Monarchies, and Republics, 1795–1871

In 1795 the First Republic assumed control with a government by directory. The directory proved to be corrupt and lacked executive power. As an alternative to the directory, the people could not resist the seduction of an ambitious military hero, Napoleon Bonaparte. France traded political ineptitude for conquest in this homegrown military genius. Napoleon's coup in 1799 created the First Empire. Seven decades of political experimentation had begun. France had seemed to lack a national sense of itself. Variations of the old and the new politics were to be tried. Emperor Napoleon's coup was just one more lurch. His military adventures ravaged Europe until he was defeated, deposed, and exiled in 1815.

In 1815 a consortium of the nations that had defeated France attempted to restore the legitimacy of the monarchy by installing the brother of the deposed and executed Louis XVI. A war-weary France accepted the restoration of King Louis XVIII, who reigned from 1815 to 1824. When Charles X, who succeeded Louis XVIII, began to behave like a prerevolutionary monarch, he was deposed by republicans dead set against a strong monarch.

In 1830 a new constitution was drafted, which brought France back to the original objective of the revolution, a limited monarchy. Louis Philippe was installed as king. He reigned from 1830 to 1848, when latent republican opposition to any monarchy gave rise to more political unrest.

The return to republican democracy came with the Second Republic under President Louis Napoleon, the nephew of the former emperor, who ran France autocratically. After four years as president, he discarded the pretense of democratic government. In 1852 he declared the Second Empire and debuted anew as Emperor Napoleon III.

His empire lasted eighteen years, ending in 1870 with the loss of the Franco-Prussian War. The German dictator Bismarck had snookered Napoleon III into declaring war on Prussia.[55] Bismarck wanted war with France to solidify nationalistic spirit among neighboring German states. His objective was to encourage Pan-German solidarity and diminish Austrian influence. It worked perfectly. In the Treaty of Frankfurt of 1871, France ceded Alsace-Lorraine to Prussia and paid a heavy indemnity.

The Third Republic was created in 1871, and France continued as a republic but with ongoing political strife between traditionalists and republicans. The strong polarization on political objectives lasted into the next century. The Third Republic survived through World War I and expired when France was overrun by Hitler's blitzkrieg in 1940 at the onset of World War II.

It had taken France almost eighty years to begin to exorcise the demons that were loosed by the revolution of 1789 and to create a semblance of stability. The process had reached some level of order in 1871, when the Third Republic was established—just before Alexis Carrel was born. Alexis Carrel was born in 1873 at the beginning of the Third Republic and died in Paris in 1944 just after the Allies liberated Paris.

Science in Carrel's Time

By Alexis Carrel's day, the monarchy was long gone, but the French Revolution had left its mark. The rationalist materialism of Voltaire and

---

[55] Gordon, Irving L., *Review Text in World History*, Ansco School Publications, 1976, pp. 204–05.

his confreres infested the universities in France. These *philosophes*, philosophers in name only, had singularly failed to fashion a humanistic path to a practical, stable democratic republic. Instead, France was treated to eighty years of lurching through seven different governments—republics, monarchies, empires. They also had their wars with their onerous costs. Carrel was born in 1873, shortly after the Franco-Prussian War ended. He died in 1944, shortly after the Allies had liberated Paris.

Throughout his life, he was unable to reconcile his science with his Catholic faith. His exclusive commitment to the experimental method was confounded by his religious sense. He had an unspecified social or political enmity toward Catholics and their Church but not Christ. He went to great lengths to bridge this divide scientifically. But at his death he was reconciled.

The French could not change their history. Their elites just chose to ignore it. Secular materialism, in Carrel's day, was the orthodoxy in the universities. What did not suit the prevailing orthodoxy might as well never have happened.

Religious or transcendent considerations of any kind were trivialized. Science appeared to be the only effective domain of reason. But today Descartes's grounding of science is known to be inadequate. The scholastic metaphysics is still there, and it continues to provide confidence in the reality of the physical world and our ability to understand it to the extent of our human limitations.

P.S. The Light of the World

About fourteen years before the Reign of Terror, on July 4, 1776, the American Colonies dealt with the king of England's oppressive taxation as follows:

The Declaration of Independence

When in the course of human events, it becomes necessary for one people to dissolve the political bands which have connected them with another, and to assume among the powers of the earth, the separate and equal stations to which the laws of nature and nature's God entitle them, a decent respect for the opinions of mankind requires that they should declare the causes which impel them to separation.

We hold these truths to be self-evident; that all men are created equal, that they are endowed by their Creator with certain unalienable rights, that among these are *Life, Liberty and the Pursuit of Happiness.*

That to secure these rights, governments are instituted among men, deriving their just powers from the consent of the governed.

That whenever any form of government becomes destructive of these ends, it is the right of the people to alter or to abolish it; and to institute new government, having its foundation on such principles and organizing its powers in such form, as to them shall seem most likely to effect the safety and happiness. Prudence indeed will dictate that governments long established should not be changed for light and transient causes, and accordingly all experience hath shown that mankind are more disposed to suffer, while evils are sufferable, than to right themselves by abolishing the forms to which they are accustomed. But when a long train of abuses and usurpations pursuing invariably the same object evinces a design to reduce them under absolute despotism, it is their right, it is their duty to throw off such government and to provide new guards for their future security.

Such has been the patient sufferance of these Colonies and such is now the necessity which constrains them to alter their former systems of government. The history of the present king of Great Britain is a history of repeated injustices and usurpations, all having in direct object the establishment of an absolute tyranny over these States. To prove this, let facts be submitted to a candid world.

May God continue to bless us.

Chapter 4

A Brief History of Physics

Nineteenth-Century (Classical) Physics
Advances in Mechanics

The development of physics is a story about people and their ideas, and ultimately their culture. Every physical law has an author. To begin, try to get a historical sense—a personal sense. You can visit and revisit any complexity later with the help of the footnotes or appendices. Isaac Newton (1642–1727) started a magnificent parade, laying the foundation of mechanics with his three laws of motion and the calculus. He gave his era a complete set of the tools needed to understand the astronomy of the solar system, plus almost all the mechanical problems on earth.

Others, such as his contemporary Wilhelm von Leibniz (1646–1716), had independently developed the calculus and had worked on a system of mechanics using the concept of energy rather than force. Force is a vector. It is geometric, having a magnitude and a direction. Energy is a simpler concept, having a magnitude only.

Following Leibniz, Johann Bernoulli (1667–1748) and Jean LeRond D'Alembert (1717–1783) used a technique called virtual work to specify the condition of a body in stable equilibrium. Think of a marble at the bottom of a bowl. If the marble is disturbed, it will move up the side of the bowl and fall right back. It is trapped in a stable state. Joseph Louis Lagrange (1736–1813) used this idea to create an energy-based alternative to Newton's laws.

In 1834 William Rowan Hamilton (1805–1865) expanded upon Lagrange's work with a more powerful energy-based theory. Hamilton's work has primacy of place in energy-based theoretical physics. These methods[56] are called analytical mechanics or analytical dynamics. The principles arising from mechanics still inform theoretical physics. Newton maintained uncontested influence until Einstein came along in the early twentieth century.

MICHAEL FARADAY

In 1813 Sir Humphrey Davy, director of the prestigious Royal Institution in London, hired Michael Faraday (1791–1867), a twenty-two-year-old bookbinder, as a lab assistant at the Institution. Faraday's hiring was due to his reputation for studious habits and copious, detailed note-taking, and for his persistent and notable presence on the fringes of the scientific community.

By 1831 Faraday the lab assistant had become Faraday the director. He had served the laboratory spectacularly well for eighteen years, maturing into a highly competent experimentalist in several areas of chemistry and physics. He had become a scientist of the first order, while providing leadership, service, and workmanship in many areas of chemistry and physics. He was poised to pursue work on electricity. Sir Humphrey Davy is said to have admitted that his most important scientific discovery was Michael Faraday.

Faraday had grown up poor with little formal education. He was literate but had poor diction and speech habits, which, in time, he took steps to correct. He stood out in ways that endeared him to all who came to know him. He was diligent and reliable. He was self-reliant with an innate integrity and a determination to learn. He was a conscientious member of a close-knit Christian group, the Sandemanians, in London.

As a journeyman bookbinder, he had access to all sorts of books. He was drawn to, and very quickly decided to focus on, science. He applied himself

---

[56] Fowles and Cassiday *Analytical Mechanics,* 6th Ed., Saunders College Publishing, 1999, pp. 391-2

to several problems of interest and fixed upon electricity. He invested his time and meager resources to develop experiments on his own, taking one step at a time as his means and opportunity allowed. He availed himself of public resources, lectures primarily, to grow in understanding. As he expressed and demonstrated interest, people were drawn to help with access and opportunities. Davy's hiring him came about through his dogged persistence in study and the reputation he gained for his virtues. He was committed, skilled, determined, and virtuous. He was hired as a lab assistant, much as an office boy in a business. But he could learn, and did, with spectacular results. After eighteen years he was appointed director of the laboratory.

Basics of Electricity

By the time Faraday took over as director, significant progress had been made in the understanding of electricity. Charge is the basic electrical characteristic of matter. Charge comes in two varieties, called plus (+) and minus (−). In France in 1785, Charles Coulomb experimentally determined the law describing the force between two electrical charges. Coulomb's law has the same form as Newton's law of universal gravitation.

The force between two charges of magnitudes Q and q, separated by a distance, d, is proportional to $[Qq/d^2]$, the same inverse square form as gravitational force.[57] The force is attractive if the charges differ in sign and repulsive if the same.

The concept of a gravitational field and field lines introduced in the section on Isaac Newton's physics and mathematics in chapter 2 applies to electrical charges and forces. The common inverse square form of the two forces suggests attraction or repulsion along radially directed lines. The

---

[57] The constant of proportionality is a number, K, which had been measured. Coulomb's law is given by $F = K[Qq/d^2]$. In addition, the electric field, E, due to the charge Q is given by $E = K[Q/d^2]$, similar in form and significance to the gravitational field. Compare to the discussion of the gravitational force and field in Chapter 2

electric and gravitational fields are represented by radial lines. See Figure 1, Lines of Force.

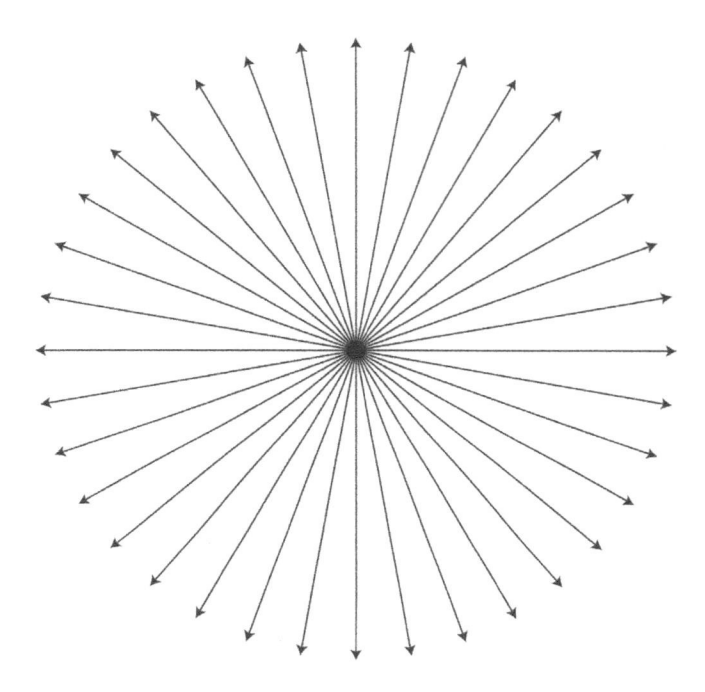

Figure 1, Lines of Force

There is a difference in scale between electrical forces and gravitational forces. Electrical charges are of atomic dimensions, while the masses considered in gravitation are of astronomical dimensions. So while there is a striking similarity in the force laws, the experimental problems are quite different. In addition, we will see that moving electrical charges are responsible for another force, magnetism.

André-Marie Ampere (1775–1836) found that a flow of electrical charge in a wire, commonly called a current, will exert a force on an adjacent wire carrying a current, evidencing that currents create forces between wires that carry them. Currents are measured in amperes. An ampere has units of charge/sec. Electric charge is not only responsible for the force between two stationary charges but also for interaction between moving charges, be they currents in wires, as Ampere found, or charges freely moving in space. This force between moving charges or wires carrying currents is a magnetic effect.

Magnetism

The first linkage of electricity to magnetism was observed in the laboratory of a Dane, Hans Christian Öersted, during a lecture to students in the fall of 1819. They observed that current in a wire will deflect a compass needle. After extensive study of the phenomenon, it was published in July 1820.[58] Öersted's investigation of the effect showed that the magnetic deflection of the compass needle tended to curl around the current in one direction only—encircling the current.

To get some picture of this phenomenon, try the following: While standing or sitting erect, extend your right arm out to the right and bend your elbow ninety degrees toward the front. Now, point your thumb to the left and curl your knuckles down. Imagine a current in the direction of your thumb. Your knuckles curl circularly in the direction of the magnetic needle's deflection.

Figure 2 shows the interaction between a current, I, and the magnetic field it creates. The magnetic field, B, is curled around, with some radius, r, and the current is directed out of the paper, toward the reader. This relationship is consistent with your right hand's orientation.

---

[58] Glashow, *From Alchemy to Quarks,* Brooks/Cole Publishing Co. Chapter 8, Independence, KY, 1994.

B - Magnetic Flux
I - Current (out)
r - Radius

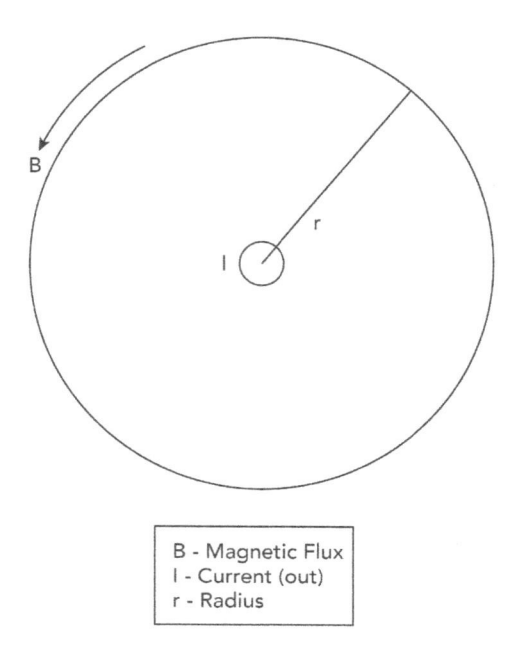

Figure 2, Right Hand Rule Diagram

Static charges cause electrical forces, and moving charges cause magnetic forces. Getting to the bottom of the relationship became the work of Michael Faraday.

Electric Force

Electric and magnetic forces have to be treated with some care. One cannot avoid geometry. As magnetism emerges, so do ideas such as flux and flux density. This account and the associated diagrams will hopefully help. Take one idea at a time.

Faraday knew little mathematics, but he was gifted with marvelous intuition. A biography[59] by James Hamilton focuses on the artist's sensibility in Faraday's science. Faraday conducted experiments convinced that there

---

[59] Hamilton, James, *A Life of Discovery, Michael Faraday, Giant of the Scientific Revolution,* Random House, New York, 2002

was a tangible "something" underlying what he saw. He was sure that every phenomenon was due to a knowable reality present before him. His job was to describe that reality.

John Tyndall,[60] who succeeded Faraday at the Royal Institution, comments on Faraday's uniqueness:

> It is impossible to say how a certain amount of mathematical training would have affected his work. We cannot say... whether it would have daunted him, and prevented him from driving his adits (mine entrances) into places where no theory pointed to a lode. If so, then we may rejoice that this strong delver at the mine of natural knowledge was left free to wield his mattock (pick) in his own way.[5]

FARADAY'S MODELS

Electric Flux Density

Faraday's models of electromagnetic reality proved to be specific, tangible, and at one with theory. His totally intuitive physical descriptions coincide with the finished mathematics of electromagnetic theory.

Faraday knew, from Coulomb, that charges exert forces on each other much as the earth and the moon attract each other. Faraday saw the space around a positive (+) charge being full of radially directed "lines of force." (See Figure 1.) If another positive charge were in the area, it would experience a force that would repel it along a line of force. A negative charge would be attracted.

The radial lines of force emanating from a central charge, Q, represent the path a positive charge at any point in the area would take. The lines are closer together (higher density) near the center, indicating a stronger force.

---

[60] Tyndall, John, *Faraday as a Discoverer,* 5th Ed., Thomas Y. Cornwell Co., New York, 1961, pp. 171-2

Conversely, the lines have a lower density at a greater distance from the center, indicating a weaker force farther away from the center.

One can visualize *electric flux* as a local group of field lines (lines of force) running near each other, in slightly different directions, like flower stalks in a vase. Just as the flowers in a vase are fixed in number, so an element of flux has a fixed number of field lines. Essentially, electric flux is a bunch of lines of force. If you select a group of lines and follow them through a distance, you are looking at a *tube* of electric flux. The tube gets wider or narrower as the lines spread out or come closer, respectively.

Faraday gave the number of lines per unit area at a point in an electric field, the *electric flux density,* a quantitative physical interpretation. The force on a charge at a point in space depends upon the *electric flux density* in the neighborhood of the charge—the higher the flux density, the higher the force.

Faraday's conceptualizations—electric field, lines of force, electric flux, and electric flux density—are basic in physics instruction to this day because they work, having meaning absolutely compatible with mathematical descriptions. These tangible representations of physical reality attest to Faraday's unique talent as a scientist.

Magnetic Force

Magnetic forces are different from electric forces in two ways:

(1) Magnetic forces are generated by charges that are moving, either as free charges moving through space or as a current through a wire.

(2) There are no magnetic charges. The familiar bar magnet is a so-called ferromagnetic material. In such a material we think of magnetic field lines arising from dipoles, combinations of coexistent (+) and

(–) magnetic "poles." A bar magnet has a north (+) pole and a south (–) pole, and you cannot have one without the other.[61]

These differences do not affect the concepts of field lines, flux, or flux density. We can think of magnetic field lines, magnetic flux, and magnetic flux density in the same way as in the electric case. Magnetic field lines, however, start on the (+) poles and terminate on (–) poles. They are not directed radially. The lines from bar magnets originate at the (+) pole and loop around to terminate on a nearby opposite (–) pole.

Electromagnetic Induction

Faraday's law of electromagnetic induction states that a changing magnetic flux perpendicular to the plane of a loop of wire will induce a current in the wire. The magnitude of the current induced is proportional to how fast the magnetic flux is changing through the loop. Similarly, a current in a wire will induce a magnetic flux perpendicular to the plane of the wire. The Ampere-Öersted observation that moving charges create a magnetic field finds its analog in Faraday's law that a moving magnet creates an electric field.

To illustrate *electromagnetic induction*, consider Figures 3(a) and 3(b). Imagine a circular metal ring like a bangle bracelet. Now, take a bar magnet and poke its north pole toward the center of the plane of the bangle. An electric current, I, will be induced within the bangle and will circulate around it. The current will circulate toward you from the top. The faster you move the magnet from the right, the greater the current induced. The moving magnet

---

[61] In fact, all materials have a magnetic character, due to the dynamics of electrons and nuclei within their atoms and molecules. Ferromagnetism is the strongest type of magnetic effect and is limited to specific combinations of chemical elements and their compounds. A bar magnet is a common example of a ferromagnet.

is changing the magnetic flux through the bangle. The changing magnetic flux produces a current in the loop.[62]

In Figure 3(b) we show what happens when we set up a battery (not shown) to create a current in the bangle. The current in the bangle induces magnetic flux through the plane of its loop. Again, the direction of the flux density depends on the direction of the current. If you create a current in a loop of wire, the current will create magnetic flux through the loop perpendicular to the plane of the loop.

To fully describe these couplings would require a geometric version of the calculus, called vector analysis. But in simplest terms electric currents induce magnetic forces and, alternatively, moving magnets produce electrical forces.

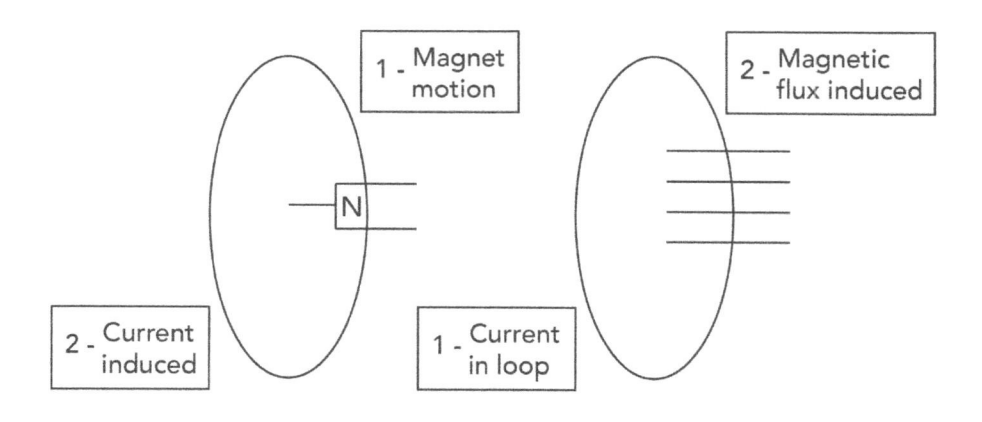

Figure 3, Electromagnetic Induction

---

[62] In this example, we have shown the north pole of the magnet approaching the plane of the bangle from the right. If the South Pole were approaching the bangle from the right, the current would circulate in the opposite direction. One can wire up many variations.

Maxwell's Electromagnetic Waves

In 1864 the Scottish physicist James Clerk Maxwell (1831–1879) published his comprehensive theory of electromagnetism. Maxwell's set of four differential equations contains all there is to know about electric and magnetic forces and the coupling between them. Faraday's law of electromagnetic induction, written as Faraday could never have written, is one of the set of four. Maxwell depended upon the mathematics of Carl Friedrich Gauss (1777–1855), a contemporary of Faraday.

Two of Maxwell's equations are expressed in terms of Gauss's law—one for electrostatics and the other for magnetism. Faraday's law of electromagnetic induction is the third. The fourth equation is a version of Ampere's Law.

Gauss's mathematics include operations analogous to Faraday's concepts of lines of force, flux, flux density, and rate of change of flux. Gauss's mathematics enabled Maxwell to summarize electromagnetism in terms of Faraday's experimentally developed ideas. To this day Faraday's law and his physical descriptions are essential tools in teaching electromagnetism.

A very important consequence of Maxwell's equations is that they predict an independent bonus, the existence of electromagnetic waves. Furthermore, the equations predict that the speed of electromagnetic waves is the same as the speed of light.

In 1888 the German physicist Heinrich Hertz produced electromagnetic waves in the laboratory. Hertz's demonstration validated Maxwell's assertion that light is an electromagnetic wave. The unexpected bonus is that Maxwell's equations not only unite electricity and magnetism but also include the understanding of optics.

Waves, unlike particles, are periodic disturbances (cycles) that travel in space with a fixed speed. Every wave has a characteristic length (wavelength) and a characteristic time (period). The wavelength, λ (lambda), of a wave is the distance occupied by one full cycle—for example, the separation between adjacent peaks. Wavelengths are measured in meters.

The period of the wave, T, is the time that it takes for a full wavelength (cycle) to pass a fixed point. Periods are measured in seconds. The velocity of a wave is the distance traveled divided by the time elapsed, λ/T.

Scientists prefer to express the time characteristic of a wave in terms of frequency, f. The frequency, f, is the reciprocal of the period, $1/T$, and is measured in Hertz (Hz), cycles/sec. A wave with a frequency of 100 Hz moves a distance of one hundred wavelengths in one second. The frequency is the number of cycles that pass a point in one second. The velocity, v, of a wave is equal to the wavelength, λ, times the frequency, f; that is, $v = \lambda f$.

Electromagnetic wavelengths range from long waves, measured in megameters, ten[6] meters, down to gamma rays, measured in femtometers, ten[15] meters. The main spectral types, in order of increasing frequency, are: long waves, radio waves, infrared (heat), light, ultraviolet rays, X-rays, and gamma (cosmic) rays. Figure 4 shows these regions of the electromagnetic spectrum wavelengths (meters, m) and corresponding frequencies (hertz, Hz).

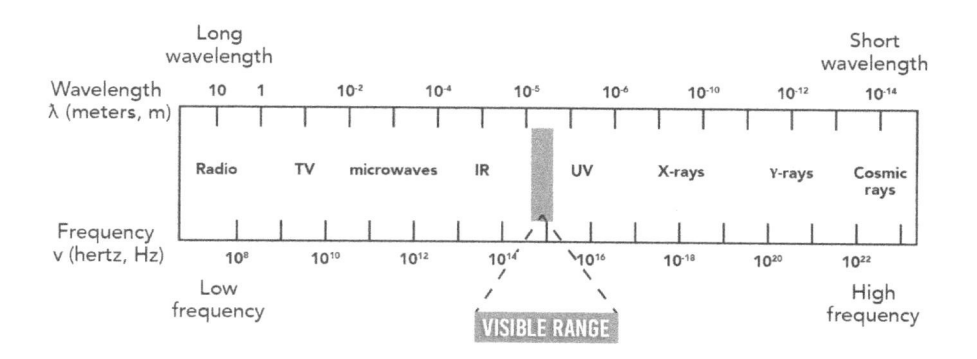

Figure 4, The Electromagnetic Spectrum

Light is electromagnetic radiation to which the human eye is sensitive, the visible range. Light occupies a very small wavelength niche from seven hundred nanometers (red) to four hundred nanometers (blue). One nanometer is a billionth of a meter. The spectrum of light is bracketed by the infrared on the high-wavelength (low-frequency) side and the ultraviolet on the low-wavelength (high-frequency) side. Cosmic rays are very high-frequency waves that originate in the universe.

Electromagnetic theory unifies electricity, magnetism, and optics. The theory is complex yet simple in form, with symmetries and asymmetries that lead to deeper insights. Electric power, telecommunications, computers, audio, video, medical diagnostics, etc. all spring from Maxwell's four equations.

The electromagnetic waves that Heinrich Hertz produced in his laboratory had to travel at the speed of light. Noting this, consider what Galileo had to say about the relative motion of waves. Galileo recognized that motion relative to a moving wave would increase the wave's speed as the observer moved toward the advancing wave or decrease the wave's speed as the observer moved away from the advancing wave. This principle is known as Galilean relativity. If you are running toward a fire hose, the water is coming faster than if you are running away from the fire hose.

Galileo would insist that a constant speed for light implies a universal reference frame, with respect to which light must move. The expected reference frame, the so called ether, was assumed to exist.

Wrong! In 1887 two American physicists, Albert A. Michelson and Edward Morley, failed to detect the ether using methods that were capable of seeing the effect if it existed. There was no ether. This anomaly, that the ether did not exist, led to Einstein's theories of special and general relativity.

Thermodynamics and Statistical Mechanics

Another major revolution in physics arose from the study of collective matter such as gasses and liquids. Thermodynamics produced fresh insights into thermal processes, the physics of heat. When thermometry, the art and science of measuring temperature, was standardized, it led to the understanding of temperature as a measure of the heat (energy) in a body.

Some physicists, following the lead of chemists, were convinced that matter consisted ultimately of small particles called atoms—an ancient idea but still hotly disputed. Atomists proposed that all matter consists of atoms or chemical combinations of atoms, called molecules.

Austrian Ludwig Boltzmann and like-minded physicists did pioneering work in applying the laws of mechanics to atoms and molecules, imputing kinetic energy to the motion of these particles. Heat was the manifestation of this internal kinetic energy of molecules, and temperature was a measure of the heat in a body. Heat, molecular internal energy, would always flow from high to low temperature regions.

Boltzmann used statistical methods to study the behavior of large aggregates of molecules—ensembles, he called them. He was able to calculate the relevant mechanical properties such as the average energy in a molecule, which he showed was proportional to the temperature. He was a pioneer in the application of the laws of motion to large ensembles of particles, a science that came to be called *statistical mechanics*.

In an ensemble, however, there is a new factor present besides Newtonian mechanics. This factor is the random nature of molecular motion. In a large ensemble of small particles, the law of conservation of energy had to apply in a special way because there was no way that the random motion of such particles could be limited in their interactions with their surroundings. Such a system was not isolated. Always some of the energy of the original system was going to be lost to the environment. Boltzmann called this element of energy "loss" entropy. Entropy is not energy lost but energy that becomes unavailable. Essentially, thermodynamic processes are not reversible, whereas in typical Newtonian systems, the mechanics of a body are reversible.

For example, the laws of physics do not forbid the reversal of a ball dropping to the ground. If a ball falls from a height, h, and hits the ground with a velocity, v, there is no problem with throwing it back up with an initial velocity, v, to attain the height, h. The physics are reversible. But once the cork pops out of the champagne bottle, the process cannot be reversed.

Boltzmann physics, published during the 1870s, were widely disputed. Sadly for him, he was too far ahead of his colleagues. It is possible that discouragement led him to take his own life in 1906.

TWENTIETH-CENTURY (MODERN) PHYSICS

MAX PLANCK AND THE BLACK BODY PROBLEM

In 1900 Max Planck, a German theorist, was trying to understand the spectrum of heat waves emitted by a black body. All of the radiation from a black body comes from within the body—there is no reflected radiation.

To make a black body takes a little ingenuity. Imagine a block of metal with a hollow center, which is maintained at a constant temperature. A hole is drilled through the block into the hollow area. The radiation that emerges from the hole to the outside is black body radiation. All of the radiation comes from within the body. The radiation is due entirely to the energy within the central cavity at the temperature of the block. If the metal is maintained at a constant temperature, the electromagnetic waves emitted through the hole have the frequency distribution (spectrum) of black body radiation at the temperature of the block.

Using existing ideas of particle energy, Planck found that the radiation emitted from a black body in the ultraviolet (beyond the violet) range was a total mismatch to his theory.

Planck sought for any means to bring the standard thermodynamic equations into line with his data. He tried all sorts of gimmicks to get his equations to match the data. But by placing a simple arbitrary restriction on the radiation energy, he was able to match the experimental spectrum exactly. Planck postulated that the energy of any black body emission had to be a multiple of a constant, 'h',[63] times the frequency, f, of the radiation. The smallest energy assignable to an atomic oscillation was hf. The energy assignment could be increased in steps of hf any number of times. Planck found that if he restricted the energy of atomic oscillators to an integral multiple of this quantum of energy, that is one, two, three . . . times hf, he could match the data.

This limitation of energy to an integer, n, times hf became the bridge between classical physics and the modern physics called quantum mechanics. The constant, h, Planck's constant, is now a principal constant of physics, like the speed of light and the gravitational constant.

All allowed energy quanta have the form nhf. Although Planck had no physical justification for this constraint, further experiments were completely in accord with it. A quantum of energy, hf, is now called a photon. Pragmatically, Planck's surmise did the trick. This quantization of energy anomaly, and the no-show ether anomaly, were on the table for the genius of the obscure Albert Einstein.

Albert Einstein's Miracle Year

There are many good books on the life and works of Albert Einstein suitable for the lay reader. Do yourself a favor: read one. Einstein was a bold and radical thinker who appreciated the importance of philosophy in grounding physics. What is the reality that underlies the observation? He was involved in a dialogue on the main lines of philosophic thought in his day. He had an intuitive sense of what was physically reasonable. Truth had to be, in some sense, simple but also beautiful. Ugly can't be true. Although he was not a practicing Jew, he believed in God based on the order in the universe.

---

[63] Planck's constant, h = 6.626 X 10^{-34} Joule-second. It is measured in units of energy times time. The type of quantity is known in mechanics as "action". Planck's constant is often referred to as the quantum of action.

As a student Einstein did not bother much with the formal requirements of his physics education. He was well informed on current physics. He wanted to practice at the frontiers of physics and was impatient with physics curricula. He studied independently, working in depth on what he considered the important and interesting problems in physics.

He left his native Germany hoping to find a more suitable environment in Switzerland. Eventually he became a citizen of Switzerland. At the Zurich Polytechnic, which was a good but not a preeminent institution, he came across as a loner. He did not endear himself to the faculty. As he neared completion of his doctorate, several proposed theses were rejected. His prospects for academic employment were slim. In 1901 a friend arranged a position for him as a patent examiner in the Swiss Patent Office in Bern while he finished his graduate work.

He was still at the patent office in 1905, the year that was his "miracle year." In 1905 he published five papers of major importance. Perhaps the least important was his April 1905 thesis on the determination of the size of molecules, which earned him his doctorate.

Planck's Quantum and the Photoelectric Effect

A month earlier in March 1905, Einstein published an analysis of the photoelectric effect. This phenomenon, which was not understood at the time, is a process in which light, incident on a metal, drives an electron out of the metal. Obviously, the light interacts with the electron within the metal to kick it out. Einstein was aware of Planck's work on a "quantum of energy" in infrared waves. He showed convincingly that, in this case, light behaves not as a wave but as a particle. He called the light particle a photon. Following Planck's work on black body radiation, he assigned a quantum of energy, hf, to the photon.

Einstein's photoelectric effect analysis showed how energy was exchanged between a photon and an electron in a metal. When the photon(s) in a beam of light enter the metal, they give one or more (n) quanta of energy, nhf, to the electron. (A quantum of energy is hf; n quanta, plural, have an energy nhf.) If the photon energy is sufficient, the electron will escape from the metal. Whatever energy was not spent in the escape remained as kinetic energy of the emitted electron. Einstein used the law of conservation of energy to determine the energy needed to release the electron, the so-called work function, and the velocity with which the electron emerged. His analysis matched the experiment perfectly. Einstein won his Nobel Prize in 1921 for this work. This work solidified the quantum theory of matter.

In May 1905 Einstein published another paper, which explained a phenomenon known as Brownian motion. When observed carefully, very small visible particles suspended in water can be seen to dart to and fro randomly, in a zigzag fashion. Einstein showed that this wiggly motion was due to a multiplicity of impacts on the particles by water molecules. This phenomenon had been a mystery for about eighty years. The effect can also be seen when sunlight reveals fine dust particles in a room. Often a dust particle darts spontaneously due to a collision with a molecule of nitrogen or oxygen in the air.

Space-Time and Relativity

Recall that Maxwell's equations require that the speed of light is constant even if an observer is moving relative to the light. This prediction means that your measurement of the speed of light from a flashlight would be the same whether you were speeding toward or speeding away from the flashlight. Einstein's fourth paper in June 1905 was on special relativity, which predicts the consequences of the speed of light being constant for all observers, independent of their relative speed.

Galileo had shown that motion of an observer relative to what he is observing determines the speed he will measure. His equations suggest that if you are moving forward at one foot per second in a train moving forward at two feet per second, an observer on the station platform will see you moving at three feet per second.

But Maxwell's work suggests that this is not true for the speed of light. For light, things are different. Physicists reconciled the apparent contradiction by assuming that light must travel in some fixed medium, the ether, which permeates all space. When Michelson and Morley did not detect the ether predicted by Maxwell, physics had to deal with a very troublesome anomaly.

Einstein claimed he had not known of the Michelson-Morley experiment but asserted that he simply did not accept the idea of the ether. He proceeded to work out the constant speed of light, implied for measurements made by two different observers of the same event in different reference frames, moving relative to each other with some constant speed near the speed of light.

Einstein starts his special relativity paper with two stipulations, which he requires from the circumstances of the problem.

1. The laws of physics must be the same for all observers. The constant, the speed of light for all observers, must not change established physics for any observer.

2. The speed of light is the same for every observer in an inertial reference frame.

A reference frame is inertial when it moves at constant velocity or zero velocity.

If the measured speed of light is constant, then common sense suggests something about the outcome. We measure a speed by dividing the distance traveled by the time it takes to travel that distance. Now if the speed of light is not going to change, then the measurements of time intervals and space intervals have to change in some manner that keeps the ratio of the two constant. The time intervals and the space intervals measured in the two reference frames will be different, but the ratio of the distance traveled to the time elapsed must be the same in each reference frame.

In beginning to tackle the problem, Einstein required that the two reference frames had to be moving with a fixed velocity relative to each other. In addition, he noted that one reference frame had to be the one within which the event occurs. The observations in that reference frame would not be affected by the relative motion. The observers in the other reference frame, however, would have to look into the reference frame of the experiment to see the event, and their perceptions would be affected by the relative motion.

The measurements made in the reference frame of the experiment are called the "proper" measurements. The measurements made there are not affected by the relative motion.[64]

We start with observer No. 1 on a train station platform and the other observer No. 2 on a train moving parallel to the platform at constant speed, v. Observer No. 1 can look into the train as it goes by. He looks over observer No. 2's shoulder as observer No. 2 measures the speed of light, c.

On the train, there is no relative motion between the observer and the experiment. Measurements of length and time made in the reference frame of the event, the train, are called the proper length and the proper time, respectively.

The observer on the station is moving relative to the reference frame of the event, and his measurements are going to be different. Einstein shows that a clock in the reference frame moving relative to the event speeds up. This speeding up of the clock is called *time dilation*.

Similarly, when a measurement is made by a meter stick moving relative to the event, the meter stick gets shorter. This shortening of the meter stick is called *length contraction*. These adjustments of time and space

---

[64] See Appendix, Time Dilation and Length Contraction, for an explanation of how a proper measurement, a measurement made in the reference frame of the event, differs from a measurement made in a reference frame moving at a constant speed relative to the reference frame of the event. The results will be that time is dilated, i.e., expanded, for the observer in the frame moving relative to the event, and the length is contracted.

measurements, however, maintain the requirement that each observer sees the same speed of light.

The nuts and bolts of Einstein's special relativity require more than ordinary effort to absorb. But if you wish, see the appendix for a detailed treatment of the problem.

Just as Planck had found a bridge between thermodynamics and molecular radiation, Einstein's special relativity postulates a bridge called space-time, a refinement of Newtonian space and time necessary to accommodate electromagnetic forces. Einstein was to construct another bridge, general relativity, to adapt his space-time to observations of non-inertial, i.e., accelerated, reference frames in 1916. General relativity essentially identifies the effect of mass as changing the geometry of space. The philosophical implications of these bridges are very significant.

Einstein's fifth paper in September 1905 is the paper on the most famous equation in the public mind, $E = mc^2$. This paper is implicit in the June paper on special relativity and is thought of as a corollary to the earlier work. However, whether the papers are thought of individually or as a pair, the work is of pivotal importance. The equivalence of mass and energy is a most revolutionary insight, relevant to nuclear radioactivity and the possibility of production of nuclear energy and weapons.

APPENDIX: TIME DILATION AND LENGTH CONTRACTION

Postulates of Special Relativity

Postulate 1. *The laws of physics are the same for observers in all inertial reference frames.* This postulate assumes that the laws of physics must be the same for observers moving at a constant speed, relative to each other.

Postulate 2. *The speed of light in free space has the same value in all directions and in all inertial reference frames.* This postulate asserts what electromagnetic theory requires: that light move with the same speed for all observers moving at a constant speed relative to each other.

An inertial reference frame is one which is not accelerated, one that is either at rest, such as a stopped train in a station, or moving at constant speed, such as a train passing through a station. An accelerated reference frame would be a train getting up to speed as it leaves a station or braking as it approaches a station. Within non-inertial reference frames, such as accelerating trains, one experiences a force due to the acceleration. A crash is tragic because the deceleration is enormous, as is the force on those within the train. A train at its operating speed is an inertial reference frame, and passengers feel no force.

Time Dilation[65]

In measuring distances and times, we designate a measured time interval as Δt. The Greek letter Δ (delta) refers to a change. So Δt, pronounced "delta tee," means a change in time, a time interval. The intervals of distance are designated by ΔL.

We will consider measurements of the same event made by two observers who are moving with a constant speed relative to each other. One observer, Steady Eddie, is on a train station platform and the other, Flyin' Ryan, is on a train moving parallel to the platform at a constant speed, v. Steady Eddie can look into the train as it goes by. He looks over Flyin' Ryan's shoulder as Ryan measures the speed of light, c. Diagram I shows the circumstances in the train.

Flyin' Ryan, within the train, is in the reference frame within which the event occurs. There is <u>no relative motion between the observer, Ryan, and the experiment.</u> Measurements made by an observer within the reference

---

[65] For a thorough discussion of special relativity see: Halliday, Resnick & Walker, *Fundamentals of Physics*, John Wiley & Sons, 1993. See Ch. 42. For the relativity of time see section 42-6.

frame of the event are called; proper length, ΔL_o, and proper time, Δt_o, respectively. Ryan's measurements are the proper measurements.

Steady Eddie, on the platform, is in a reference frame that is moving horizontally relative to the train. Steady Eddie's relative horizontal motion is perpendicular to the motion of the light beam, so Steady Eddie has no relative motion vertically, just horizontally. As a result, Steady Eddie sees the vertical measurement just as Flyin' Ryan does. Eddie's vertical measurements of length are the same as Ryan's.

However, the train is moving with a velocity, v, horizontally. Therefore, the light beam has a horizontal component of motion, as seen by Steady Eddie. Diagram II shows the motion of the light beam as seen by Steady Eddie on the platform.

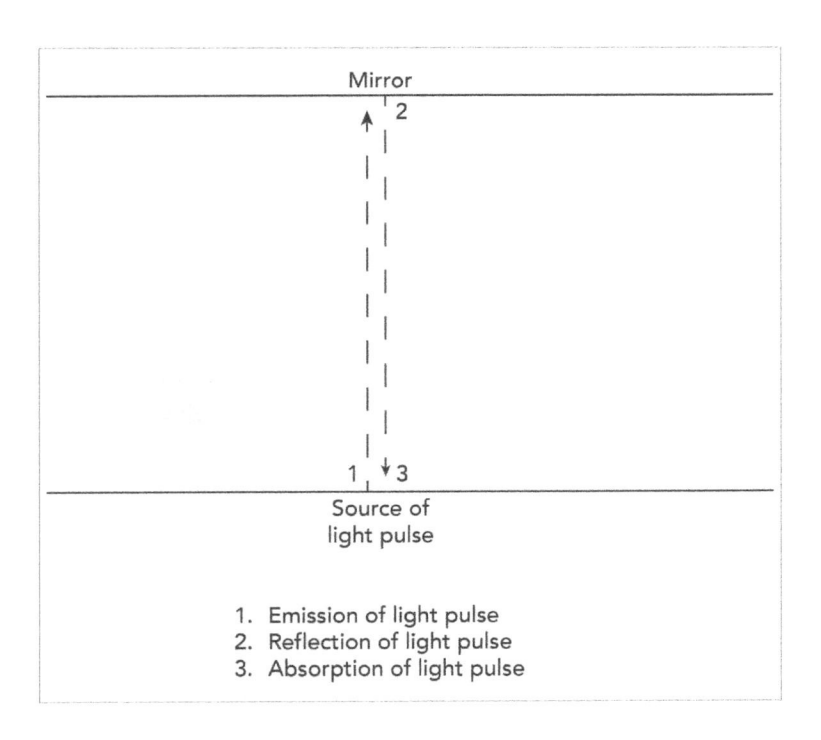

Mirror

1. Emission of light pulse
2. Reflection of light pulse
3. Absorption of light pulse

Figure 5, **Diagram I**

111

In Ryan's reference frame, a light source, S, is separated by a vertical distance from a mirror, M. The vertical distance between S and M, as measured by the meter stick, is L. At time t = 0 a clock starts when a pulse of light leaves the source, S. The pulse moves a distance, L, to the mirror and is reflected back a distance, L, to the source. When the pulse returns to the source, the clock is stopped. The measured time interval between the emission of a light pulse and its reception back at a detector at S is the proper time, Δt_o. Note Δt_o = 2L/c, where c is the speed of light and the proper length ΔL_o = 2L.

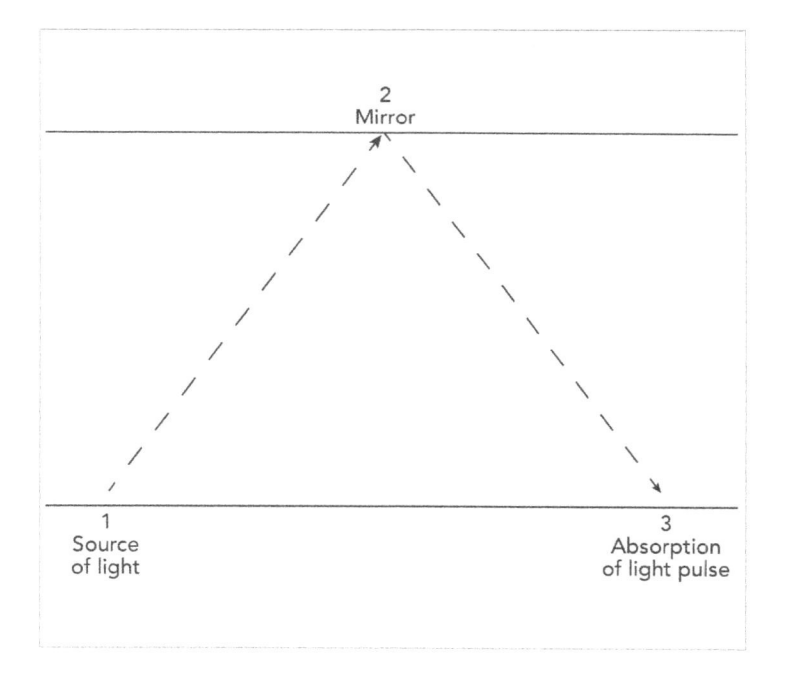

Figure 6, **Diagram II**

Let Δt be the time interval that Steady Eddie measures for the light to travel to the mirror and back. As he sees it, the light moves up a distance, $L_o/2$, and to the right a distance, $[v(\Delta t)]/2$. When the light is reflected by the mirror, it moves down a distance, $L_o/2$, and to the right a distance, $[v(\Delta t)]/2$, to reach S'. The time elapsed is detected by a second clock, synchronized with the first, on the platform at S'. We are assuming that by the symmetry of the event that the time taken to reach the mirror and the time to come back to the source is the same, $\Delta t/2$.

Note that the total distance traveled is 2D, i.e., D to go up and D to go down. The Pythagorean Theorem enables us to calculate D:

$$D^2 = (L_o/2)^2 + \{[v(\Delta t)]/2\}^2$$

Note that 2D must equal $c\Delta t$, therefore, $D = [c(\Delta t)]/2$.
We know that, $L_o/2 = [c\Delta t_o/2]$ so,

$$\{[c(\Delta t)]/2\}^2 = [c\Delta t_o/2]^2 + \{[v(\Delta t)]/2\}^2$$
$$(c^2-v^2)(\Delta t/2)^2 = [c\Delta t_o/2]^2$$

The time interval measured from the platform, Δt, is given by:

$$\Delta t = \Delta t_o/[1-(v/c)^2]^{½}$$

If we let $\beta = (v/c)$, then:

$$\Delta t = \Delta t_o/[1-\beta^2]^{½}$$

And if we let $\gamma = 1/[1-\beta^2]^{½}$, then:

$$\Delta t = \gamma \Delta t_o$$

Note: $\beta = v/c$ is a number less than one and greater than zero. Since $[1-\beta^2]^{½}$ is less than one, necessarily, $\gamma = 1/[1-\beta^2]^{½}$ is greater than one. Therefore, $\Delta t > \Delta t_o$, the time interval, Δt, measured by Steady Eddie is greater than the proper time, Δt_o, *measured by Flyin' Ryan. This increase in the time observed in the inertial reference frame moving relative to the inertial reference frame of the event is called time dilation.*[66]

This kind of explanation is of a type commonly used in the discussion of relativity, called a gedunken experiment, meaning, a "thought" experiment. We have to imagine that we can, in principle, position clocks at any point in the platform reference frame and that the clocks are all synchronized. As noted previously, for simplification, we assumed that the distance traveled by the light beam had to be equal before and after the reflection, so the two time intervals had to be the same.

Length Contraction

We will approach the phenomenon of length contraction by taking advantage of Einstein's two postulates[67] and the time dilation result that we have just established—that the time interval measured in the reference frame moving relative to the frame of the event is increased by a factor, γ, i.e., $\Delta t = \gamma \Delta t_o$.

We are going to consider a different kind of event, essentially how a clock is affected by special relativity. We are going to talk about a very short-lived particle called a muon.[68] Muons are unstable particles created in our atmo-

---

[66] Note that to measure the time interval on the platform, Eddie must use three synchronized clocks in his reference frame. The first clock has located at the position of the light source on the station platform when the light pulse is emitted by the source on the train. The second clock is located at the position of the reflecting mirror on the station platform when the light pulse is reflected by the mirror on the train. The third clock has to be located at the position the detector of the reflected light pulse on the station platform as the reflected light is absorbed by the detector on the train. Measuring a time interval which is not proper requires a pair of clocks. In this case the two time interval measurements require three clocks.

[67] 1. The laws of physics are the same for observers in all inertial (unaccelerated) reference frames.
2. The speed of light in free space has the same value in all directions and in all inertial reference frames.

[68] Beiser, *Concepts of Modern Physics,* McGraw-Hill, NY, 2003. See Section 1.4, p. 15.

sphere by the incoming radiation and particles from the universe. These high-energy particles and radiation interact with the atomic nuclei of the gasses in the atmosphere to form muons.

Physicists have studied muons in particle accelerators, and they have measured their individual properties as particles. They have formed a very good theory, the standard model, within which a muon is understood in relation to other subatomic particles. A muon is like a big electron, with the same charge as an electron, which can be either plus or minus. However, the muon's mass is more than two hundred times that of an electron. It has been found experimentally that muons travel at speeds equal to 99.8 percent of the speed of light, i.e., 0.998c or 2.99×10^8 meters per second (m/s).

The muon decays into other particles in about 2.2 microseconds (μsec). A microsecond is one-millionth of a second, 10^{-6} seconds. Because its lifetime is a known characteristic of the muon, we can consider the muon a clock that records a 2.2 μsec time interval. The muon lifetime is a proper time interval, Δt_o = 2.2 μsec in its reference frame.

Since we know the muon's lifetime, 2.2 μsec, and its speed, 2.99×10^8 m/s, we know the distance it travels in its lifetime, 660m or 0.66km. This length, of course, is the muon's proper length, ΔL_o, the distance it travels in its own reference frame during its lifetime, Δt_o.

Ordinarily one might expect that a muon could never be detected on earth, because they originate higher than six thousand meters, six kilometers above sea level. Yet we detect them. Why?

We on the earth are not in the reference frame of the event. We are looking at a clock, which is in a reference frame that is moving relative to us at a speed of 2.99×10^8 m/s. Because we know that our time is dilated relative to the proper time, Δt_o, our clock is going to tick faster than the muon clock. Our clock will record a longer time for the fall of the muon toward earth.

Recall in general that time dilation is given by $\Delta t = \gamma \Delta t_o$. Also, $\beta = v/c = 0.998$, and we know that $\gamma = 1/[1-\beta^2]^{\frac{1}{2}}$. Doing the arithmetic, we get $\gamma = 15.8$. Therefore, in the earth reference frame, the dilated time interval is 15.82 x

2.2 μsec = 34.8 μsec. So in the earthlings' reference frame, the muon has a lifetime of 34.8 μsec.

Since the relative motion of our reference frame is 2.99 x 10⁸ m/s, in 34.8 x 10⁻⁶ seconds, the muon clock will fall a distance of 10.4 km, sufficient to reach the sea level. If we compare the ratio of the earth frame clock ticks, 34.8, to the muon frame clock ticks, 2.2 clicks, we get a ratio of about 15.8 to 1. This is the same as the figure we had for γ in the time-dilation result.

According to Einstein's second postulate, the speed of light has to be the same in the earth's reference frame as it is in the muon's reference frame. Now, the speed of light in the muon's reference frame is $c = \Delta L_o/\Delta t_o$, and in the earth reference frame, $c = \Delta L/\Delta t$. Obviously, $\Delta L_o / \Delta t_o$ and $\Delta L/\Delta t$ have to equal each other. We can invoke time dilation ($\Delta t = \gamma \Delta t_o$) to figure out what relation has to exist between the proper length and the length associated with the earth reference frame. Thus:

$$\Delta L_o / \Delta t_o = \Delta L/[\gamma \Delta t_o]$$

This can be rearranged to show that the length in the earth reference frame has to be reduced by a factor, $1/\gamma$:

$$\Delta L_o / \Delta t_o = [\Delta L/\gamma]/ \Delta t_o$$

If the length measurements in the earth reference frame moving relative to the frame of the event (muon lifetime) were reduced by the same factor, γ, as the dilated time was increased, then the speed of light would be the same in both reference frames. Thus, we can say that, because the speed of light has to be the same in each reference frame, the space interval measured in a reference frame moving relative to the frame of an event must be contracted by a factor, γ.

Chapter 5

Science and Other Ways of Knowing

Pascal's Epistemology

Blaise Pascal was a master of the first order of knowledge, the order of the senses. He excelled in experimental science, particularly in studies of the vacuum and in measuring the pressure of the atmosphere. In addition, he engineered a technological marvel, the arithmetic machine.[69] On the order of the mind, Pascal was a gifted mathematician. The best-known of his works were his theory of probability and his arithmetic triangle.

Pascal had a unique epistemology, philosophy of knowledge. He developed this understanding based upon his observation of human behavior and on his own growth in the spiritual life. Scripture gave him an awareness of man's relationship to God and how God can be known.

His personal experience of the presence of Jesus Christ on November 23, 1654, gave him a unique certitude. He received an infused knowledge of being on the order of the heart. His Memorial experience dramatically proclaims this new knowledge—"Certainty, Certainty, Certainty."

St. Thomas Aquinas late in his life had a similar experience, after which he characterized all his philosophy as just "straw." Pascal found his certainty

---

[69] A replica of Pascal's arithmetic machine is on display at IBM's T. J. Watson Research Center in Yorktown Heights, N. Y.

not in the ponderous reasoning of metaphysics, true as it may be, but in the compelling two-hour Memorial experience. He knew that scripture was correct on questions of man's life and on the reality of creation. This Memorial experience, it seems, gave him a unique insight into what constitutes a proof. He proclaims in his *Pensées* that a proof is an argument or proposition that persuades.

Sometime in 1658,[70] Pascal began to organize a collection of notes, essays, comments, and fragments he had written in preparation for an *Apologia* on the truth of the Christian faith. In the spring of that year, he had indicated to friends at Port Royal his intention to publish such a treatise. These *Pensées* (*Thoughts*) detail how the Christian faith fulfills the Old Testament prophesies of the Messiah and how Jesus Christ, the Redeemer, fully respects the freedom of man while addressing his wretchedness due to the fall. This large collection of notes of varying length—from mere phrases and sentences to lengthy essays—was organized by Pascal into broad areas, obviously as preparation for writing.

Pascal argues that true religion must address and reconcile two contradictory truths that color man's experience of himself—his greatness as well as his wretchedness. The true religion must explain these contradictions in man—his dark sense of himself along with his utopian striving. It must address his brokenness as a being of contraries. Saint Paul, in his *Epistle to the Romans* (Ch. 7, 18–19), describes this wretchedness in the midst of his striving for righteousness: "I know that good does not live in me, that is, in my human nature. For even though I desire to do the good, I am unable to do it. I don't do the good I want to do; instead, I do the evil that I do not want to do." The spirit is willing, but the flesh is weak.

---

[70] Natoli, Charles M., *Fire in the Dark, Essays on Pascal's Pensées and Provinciales*, U. of Rochester Press, 2005, pp. 9-95.

Pascal's Three Orders

Science is validated by sense experience. It works for the birds and the bees, the flowers and the fleas. Their senses are akin to our knowledge on the order of the body. However, unlike other creatures, we are not just physically alive. Man is a transcendent spirit with an intellect and a will. So when a man's senses do what they do, man has an innate drive to conceptualize, to form an idea. When he conceptualizes, on the basis of sense experience, he makes sure that his idea is consistent with everything else his senses tell him. Good science requires that every sense-based idea is, to some degree, provisional—subject to further experiment. There is an ongoing interplay between the order of the senses and order of the mind.

Man's intellect, by its very nature, seeks a more general understanding, a logical framework to encapsulate everything his senses provide. He seeks an understanding that will apply to every case of whatever phenomenon he is sensing. The order of the mind propels him to find a way to grasp the principles of this particular physical behavior. The practice of physical science is a progression that seeks more complete intellectual constructs as the sense data is refined and tested. A good theory should suggest new experiments to test its validity.

Human reason comes with its own powers—inherent principles and intuitions, e.g., space, time, and number—that constitute the reach of the mind. The analytic capacity of the mind can know a real being, recognize the good in the being, and access that good. Theoretical scientists take the data of experiment, the first level, and try, through reason, to find some order in physical behavior

This reach of the mind enabled Pascal to conceive the arithmetic machine, a prototype computer. The modern computer is a physical interface between the order of the senses and the order of the mind. In order to work, the computer must be programmed. The dummy must be told what to do. When the knowledge of the mind is encoded in the device, it does what it is designed to do. If we know the mathematics and the physics are right, the computer will work while knowing nothing.

Pascal vigorously insisted on the integrity of sense knowledge in its own domain, experiment. Experimental science is the sole arbiter in settling purely physical questions, issues within the first order (the body). Such questions are settled by evidence of the senses. Pascal vigorously defended belief in the existence of a vacuum from his experiments exclusively. He resisted, as did Copernicus and Galileo, the intrusion of philosophical dispositions about physical questions.

The third order of knowledge is due to the promptings of the heart—knowledge that can be experienced only through love, most deeply love of God for us, agape, sacrificial love. There are some pointers toward this higher order—for example, the mutual knowledge of spouses, the relationship between parent and child. Also, the moral imperatives found in disparate cultures imply a higher order of knowledge.

Living between the orders of body and mind, one is hard-pressed to experience love. St. Paul says we see through a glass darkly. Augustine[71] saw his embodied spirit able to turn upward to God with love or downward toward the satisfaction of the body. He records his mind searching, facing upward toward love of God or inward to sense knowledge of the corporeal world. He called this choice the two wills.

---

[71] Augustine, *Confessions*.

Describing the Orders

Excerpts from *Pensée* No. 308[72] are selected to summarize how knowledge in a given order is isolated from knowledge in a lower or higher order. Rather than using quotation marks, the text of the *Pensées* will be *italicized*.

1. To show isolation of the higher orders from the lower orders:

   The infinite distance between body and mind symbolizes the infinitely more infinite (sic.) *distance between mind and charity, for charity is supernatural. All the splendor of greatness lacks luster for those engaged in pursuits of the mind.*

   The greatness of intellectual people is not visible to kings, rich men, captains, who are all great in the carnal sense.

   The greatness of wisdom, which is nothing if it does not come from God, is not visible to carnal or intellectual people.

2. To show the capacities of the higher order are independent of the lower:

   Great geniuses have their power, their splendor, their greatness, their victory and their luster, and do not need carnal greatness, which has no relevance for them. They are recognized not with the eyes but with the mind, and that is enough.

---

[72] The numbering of the *Pensées* has been revised as scholarship has developed. The original French is the Brunschvieg edition. The numbers used here are from the English Krailsheimer translation; first published by Penguin Books in 1966. A newly developed Sellier numbering system is also abroad and is employed in some recent scholarship. See Charles Sherrard MacKenzie, Blaise Pascal Apologist to Skeptics, University Press of America, New York, p. xix.

Saints have their power, their splendor, their victory, their luster, and do not need either carnal or intellectual greatness, which has no relevance for them, for it neither adds nor takes away anything. They are recognized by God and the angels, and not by bodies or by curious minds. God is enough for them.

Jesus' simplicity on the lower orders (body and mind) *had no effect on his infinite love (heart):*

Jesus, without wealth (body) *or any outward show of knowledge* (mind*) has his own order of holiness. He made no discoveries* (mind*); he did not reign* (body*), but he was humble, patient, thrice holy to God, terrible to devils, and without sin. With what great pomp and marvelously magnificent array he came in the eyes of the heart, which perceive wisdom!*

It would have been pointless for Our Lord Jesus Christ to come as a king with splendor in his reign of holiness, but he truly came in splendor in his own order.

It is quite absurd to be shocked at the lowliness of Jesus, as if his lowliness was of the same order as the greatness he came to reveal.

If we consider his greatness in his life, his passion, his obscurity, his death, in the way he chose his disciples, in their desertion, in his secret resurrection and all the rest, we shall see that it is so great that we have no reason to be shocked at a lowliness which has nothing to do with it.

3. To see how the lower orders can recognize greatness of the upper orders:

> *All bodies, the firmament, the stars, the earth and its kingdoms are not worth the least of minds; for it knows them all and itself too, while bodies know nothing.*
>
> *All bodies together and all minds together and all their products are not worth the least impulse of charity. This is of an infinitely superior order.*
>
> *Out of all bodies together we could not succeed in creating one little thought. It is impossible, and of a different order. Out of all bodies and minds we could not extract one impulse of true charity. It is impossible, and of a different, supernatural, order.*

Being able to hit a baseball like Babe Ruth (body) implies nothing about his mind or heart. Winning a Pulitzer Prize (mind) does not imply that the author is responsive to the promptings of the heart. The finest in a carnal order (body) confers no sense at all of the order of the mind. And similarly, the greatest of minds does not confer the majesty of a saint (heart).

As embodied spirits we have to live on the lower levels (body & mind), but the mind and the senses are not our highest capacity. Francis of Assisi's sanctity was so fulsome that he was not enthused by attempts to structure the Franciscan community. In the same way, a scholar's eminence is untarnished by his inability to hit a curveball.

Eminence in an order (body, mind, or heart) does not convert equally between orders. Mother Teresa addressing a prayer breakfast in Washington seemed to have rapt attention of the many politicians, but her message was received differently, maybe heart to heart, maybe mind to mind, or maybe by deaf ears. Einstein's eminence in intellectual achievement made his name an occasion for awe in the language of the common man.

Achieving the types of knowledge gained on the three orders are manifestations of how fallen man can be disposed to the lowest and can yet aspire to the highest. A man who decides to accept his wretchedness, who ignores his innate stirring for perfection, has made a decision. He has chosen not to pursue his vague longing, an ontological remnant of a past bliss. God has made him free. He will not compel him. Pascal's *Pensée* 149 ends by citing the freedom of the human will as the reason God hides: *There is enough light for those who wish only to see and enough darkness for those of contrary disposition.*

Pascal suggests that man's ability to see depends upon his desire to see. A man who, though living with physical and intellectual limitation, sees the dim light of love and pursues it will have some experience on the order of love. One who is ill-disposed to the light will be amply distracted by either the mind or the senses, or both.

Pascal identifies the defenses against the light as indifference to it and/or distraction from it. How commonly are the responses "Whatever!" or "Get a life" invoked against the promptings of the heart today? God will butt out, but He will not go away. His light will flicker on until one's death.

What Is a Proof?

Pascal viewed philosophical proofs for God's existence as valid logically, on the order of the mind, but lacking rhetorical compulsion—cold and dry. They do not appeal to the heart. Pascal believed that a fully alive human heart, desiring only to know, with a will opened to hearing, may be able to quell the noise of the world and hear the music of God. Such a heart, fully alive, will more readily respond when the gift of love is given. His *Pensées* reveal this broader understanding of what it means to prove something.

Consider *Pensée* 781:

> *I marvel at the boldness with which these people presume to speak of God. In addressing their arguments to unbelievers,*

their first chapter is the proof of the existence of God from the works of nature. Their enterprise would cause me no surprise if they were addressing their argument to the faithful, for those with living faith in their hearts can certainly see at once that everything which exists is entirely the work of the God they worship.

But for those in whom this light has gone out and in whom we are trying to rekindle it, people deprived of faith and grace, examining with such light as they have everything they see in nature that might lead them to this knowledge, but finding only obscurity and darkness; to tell them, I say, that they have only to look at the least thing around them and they will see in it God plainly revealed; to give them no other proof of this great and weighty matter than the course of the moon and the planets; to claim to have completed the proof with such an argument; this is giving them cause to think that the proofs of our religion are indeed feeble, and reason and experience tell me that nothing is more likely to bring it into contempt in their eyes.

Philosophical proofs do indeed resonate with the believer. But philosophical proofs do not persuade the atheist, who is stuck on the first and second level of knowledge and does not have a sense of the Creator. What will persuade him? Pascal finds the answer in the approach of Scripture, which directly addresses the unbeliever in the very isolation and confusion of his heart.

Pensée 781 continues:

This is not how Scripture speaks, with its better knowledge of the things of God. On the contrary it says that God is a hidden

God, and that since nature was corrupted, he has left men to their blindness, from which they can escape only through Jesus Christ without whom all communication with God is broken off. Neither knoweth any man the Father save the Son, and he to whomsoever the Son will reveal Him.[73] *This is what Scripture shows us when it says in so many places that those who seek God shall find him. This is not the light of which we speak as of the noonday sun. We do not say that those who seek the sun at noon or water in the sea will find it, and so it necessarily follows that the evidence of God in nature is not of this kind. It tells us elsewhere: Verily thou art a God that hidest thyself.*[74]

Pascal believes a proof may not even be an argument but a condition or circumstance that persuades the person of the truth of a proposition. Those who know God on the level of the heart have a direct sense of his Lordship as the Creator.

On a summer evening some years ago, as I walked across campus facing a beautiful sunset, I commented on its beauty to a student walking nearby. She answered from the heart, "Blessed be His holy name."

For a comprehensive discussion of Pascal's standard of rhetorical persuasion in a proof, see the essay, *"Proof in the Pensées: Reason as Rhetoric, Rhetoric as Reason."*[75]

---

[73] Mt. Ch. 11, verse. 27

[74] Is. Ch. 45, verse 15

[75] Natoli, Charles M., *Fire in the Dark, Essays on Pascal's Pensées and Provinciales,* U. or Rochester Press, 2005, pp. 69-95

Carrel at His Peak

Carrel's Religious Sense

Carrel was born into the Third Republic in 1873. Its founding in 1871 had followed a humiliating defeat in the Franco-Prussian War. It ended eighty-four years of political and military adventurism.[76] Carrel was a philosophical casualty. His religious sense is evident in his frequent animated discourses on religion—diary entries, memoirs, etc.

He refused to practice his faith, based on a loathing of people who do.[77] He had absorbed much of the skepticism of his day. He was intent on building his knowledge by inductive methods. He fully understood the limitations of the method, particularly its requirement of finding ways of measuring by use of the senses only.

He saw the possibility for the scientist studying man to overemphasize his data, a clear temptation from Cartesianism. "A phenomenon does not owe its importance to the facility with which scientific techniques can be applied to its study. It must be conceived in function, not of the observer and his method, but of the subject, the human being."[78]

He continues in the same paragraph: "The grief of a mother who has lost her child, the distress of the mystical soul plunged into the 'dark night,'

---

[76] After Louis XVI was guillotined, the constitution of the First French Republic was suspended until 1795, while National Convention and its notorious Committee on Public Safety pursued war against other European monarchs. The First French Republic was overthrown by Napoleon in 1799. He ruled the First French Empire until 1815, when exiled. He was followed as emperor by Louis XVIII and Charles X until a republican coup brought in a limited monarchy under Louis Philippe, who reigned until 1848 when a new constitution created the Second French Republic under President Louis Napoleon, who in 1852 thought Emperor Louis Napoleon III sounded better. So he created the Second French Empire. He met his demise in 1870 when he lost the Franco-Prussian War, with catastrophic consequences for France.

[77] Durkin, *Hope for Our Time*, p. 66.

[78] Alexis Carrel, *Man the Unknown*, Harper & Brothers, 1935, p. 38.

the suffering of the patient tortured by cancer, are evident realities, although they are not measurable." Here he recognizes, with Pascal, knowledge of the heart and the mind, which are beyond the reach of measurement.

Carrel, however, proposed a unique extension of the observational methods for physiologists like himself. He avowed that he and physicians like him could record physical indicators of psychological states. The emotional, affective, or moral state of a person could be reflected in his physiology. He writes: "Mental and spiritual activities, when they play an important part in our life, express themselves by a certain behavior, certain acts, a certain attitude toward our fellow man. It is only in this manner that the moral, esthetic, and mystic functions can be explored by scientific methods."[79]

The Rending within France

The historic perspective of Carrel's France had been shaken by the French Revolution and by eighty or so years of furious military adventures and convulsive politics. In addition, the absolute finality of regicide sealed a total disempowerment of the privileged elites. The Committee for Public Safety, under the fanatic Robespierre, put its bloody mark on the revolution, creating a legacy of hatred and division that would poison French politics for a very long time.[80]

Finally, when a country is transformed as radically and as brutally as France was, people do not look back with discrimination. In fact, a whole intellectual and spiritual tradition was abrogated. France sought to displace an intellectual history of almost two millennia *ad hoc*. Anything identified with the *ancien regime* was discredited. The fragile bones *"Liberté, Égalité, Fraternité"* carried little meat. The France that emerged was defined by what

---

[79] Ibid. p. 41

[80] See, for example, H. G. Wells, *The Outline of History*, Garden City Publishing, 1920, Ch. XXXVI

it was not and affected a defensive hyperpatriotism and a materialist rationalism, called the Enlightenment.

In the meantime, experimental science was doing marvelously well. Science was good news in a bleak time. It looked like a very safe mooring in the intellectual void. Science became a shining light for Carrel and many others. An incipient materialist ideology resulted, featuring a pan-scientific dream—science could address all problems.

It is striking how Carrel can speak eloquently of the dangers his own ideas contain. In commenting on pure science, he offers: "... when its fascinating beauty dominates our mind and enslaves our thoughts in the realm of inanimate matter, it becomes dangerous."[81]

EPISTEMOLOGICAL RESTRICTIONS

As a young doctor, Carrel had been torn by uncertainty—his inability to answer basic questions.[82] He agonized over life's contingencies—the certainty of death, the struggle to succeed, the suffering life brings. Was it worth it? Was he selfish? Could he ever be happy? At one time contemplating professional failure, he speculated on the possibility of being a farmer.

His angst was reasonable. There were no footprints on the path he was following. He was bedeviled by the peril of being a pioneer, a new type of medical scientist. But anxiety, while uncomfortable, never dissuaded him. He was going to study the systematics of living bodies—life processes in cells, tissues, and ultimately organs.

His experimental tools would become a major contribution to biology and physiology as a whole. He would move on to find ways to study organs outside the body and to develop mechanical replacements for organs. In truth, at the time, he was in line for the Nobel Prize based on what he had done already.

---

[81] Carrel, *Man the Unknown*, p. 42

[82] Durkin, pp. 54-58

Carrel had come of age with an ambivalent religious sense, a self-imposed epistemological standard, clearly influenced by the materialist rationalism of the scientific community. He was not a knee-jerk materialist, but he stayed as close as he could to the scientific method. He did not pursue philosophy in depth.

His strongest philosophic reaction was anti-Cartesian. He saw the Cartesian reductionism in the practice of science as inadequate to the study of human life. He proposed to augment the quantitative character of scientific study with a qualitative component, which he outlined in his 1935 bestseller *Man the Unknown*. He also tried to frame a science–oriented relation with Jesus Christ.

Christian Scientism

Though the spiritual life is opaque to scientific scrutiny, Carrel thought he could apply science to nonphysical realities—even situations involving God's love. In a diary entry written in Chicago in 1905,[83] still seeking an appropriate position, pondering medicine in South America, he writes: "Jesus Christ is the road to the truth and to life. But if loving Jesus Christ and attaching oneself to Him consists in following the way of truth and of life, it is equally true that to love the truth above all things and to consecrate oneself to it and to place all one's hope in it—even if this means disowning and denying Christ—this is even a way of following Him, marching in His steps, and finding Him."

God is not hidden here. Carrel sees and is drawn to Christ in love. He wants to reciprocate. But he is skittish about responding with love from his heart. He clearly is affected by the materialist intellectual fashion. Love is from the heart, not the intellect. He wants intellectual knowledge, not the direct, nonscientific, knowing that is love.

---

[83] Ibid. pp. 64-65

He doesn't like the shortcut God takes around his senses. It is as if God were breaking some rule. He is disposed to scientific means only. He tries to fashion an acceptable alternative. He suggests a subjective approach, one in which he is in charge.[84] He expounds: "When the veils are drawn aside, when the Great Day replaces the shadows of the present hour, those who have turned away from Christ because they never knew Him, will recognize Him by their pursuit of the truth. It is His spirit that they seek although they do not realize that fact. It seems to me that the suffocating sighs of these great spirits, who have been the sorrowful martyrs of their own sincerity, are the prayers that rise the straightest to the God of truth."

Passionate seeking of the truth, scientifically, is the only means available to access Jesus Christ for those who refuse to respond to the promptings to the heart. God's gift, His grace, is unrecognized.

The "Spirit blows where it will,"[85] and God may lift the soul by other means. But Carrel's refusal of love as an entrée to knowing Christ had to derive from a very deep source—a strong negative influence within him.

In continuing the meditation Carrel reveals the impetus for this choice: "Since it is definitely impossible for me to ignore God, and since it is impossible to remain on the horns of the dilemma, it is necessary that I believe. But all belief implies action, and here begin the difficulties that I am totally unable to surmount." Why is he incapacitated?

He reveals the source: "To return to the milieu of the feeble of spirit, of the egoists and the contemptible people who comprise Catholic society today, to try to reenter into relations with the priests who are mere honest functionaries, brave ignorant men who have transformed the Sermon on the Mount into some sugary and infantile formulas that the devout mutter in their [words missing]—all this frightens me; to return there is to risk permanent moral and intellectual asphyxia."[86]

---

[84] Durkin, p. 65

[85] Jn. 3:8

[86] Durkin, p. 66

Undoubtedly, but what motivates his anger is unclear. For some reason he disdains practicing French Catholics. He identifies the Catholic Church with people who are for one reason or another abhorrent. He will find Christ his own way. There is no failed epistemology here, just a deep resentment—maybe personal experience, maybe a social injustice. He will maintain his point of view almost to the end of his life. His personal diaries throughout his life show that he prayed and believed in God, but he did not feel compelled to respond to Christ's love heart to heart.

The Study of Miracles

Carrel's interest in the cures taking place at Lourdes was evident six months before his witnessing the cure of Marie-Louise Bailly. In a diary entry on December 8, 1901, Carrel muses: "If God wants me entirely, I shall rush toward him. Without stint I shall consecrate myself to his service." He then reveals his interest in miracles: "I shall pass these months in studying and digging deep into the question that chiefly interests me—the question of miracles."[87]

His involvement in the cure of Marie-Louise Bailly was a challenge to his Christian scientism. Miracles are manifestations of the divine in the physical world and constitute a problem for the primacy of scientific reasoning. In his later writings, when appropriate, he never fails to mention the events at Lourdes and appears, as he ages, to be stronger in his tolerance of, if not belief in, the existence of miracles.

In his *The Voyage to Lourdes,* Carrel describes the thoughts of his alter ego, Lerrac, after witnessing the cure of Marie Ferrand, the name given to Marie-Louise Bailly in the book:

> Marie Ferrand was a 'miracle cure.' Here was a girl on the point of death at noon and well again by seven in the evening. Such a thing justified an outburst of popular fervor.

---

[87] Durkin, p. 58

But deep in his own mind, what was he to think? Profoundly uncertain, he pondered the only two possibilities: Either he had made a grave error in diagnosis, mistaking nervous symptoms for an organic infection, or else tubercular peritonitis had actually been cured. Either he had made a mistake, or seen a miracle. His mind rushed on to the inevitable question: What was a miracle? [88]

Carrel was deeply conditioned by his philosophy. A few short pages later, he laments: When a scientist tried to apply his intellectual techniques and convictions to metaphysics, he was lost." Metaphysics was a catchword for anything he could not understand. "He could no longer use his reasoning, since reason did not go beyond the establishing of facts and their relations to each other. In search for causes, there was nothing absolute. There were no signposts along the way. There was no proof of right or wrong. All things in this mysterious realm were therefore possible."[89]

Carrel is convinced that unless he misdiagnosed Mlle. Bailly, a miracle had occurred. It is a tribute to his integrity that he fully recorded his observations to the medical board at Lourdes, somewhat begrudgingly but honestly. He knew that his rationalist colleagues would hold him in contempt for being involved with Lourdes. Yet he maintained interest in Marie-Louise Bailly until her death at age fifty-eight.

In his book, *Prayer,* written in 1942, he refers to events at Lourdes as being a consequence of prayer. He makes specific reference to the medical board at Lourdes, commenting that their data "has rendered a great service to science in demonstrating the reality of the cures."[90]

In 1913 Carrel married Anne-Marie de la Motte de Meyrie, a widow with one son. He had seen her at Lourdes on his third visit to the shrine

---

[88] Carrel, *The Voyage to Lourdes,* Real-View-Books, 2007, p.86
[89] Ibid: pp 92-3
[90] Carrel, *Man the Unknown,* p. 42

in 1910. She was a volunteer nurse at the shrine. On that occasion Carrel had a second experience of a spontaneous cure. An eighteen-month-old boy, born blind, was in the arms of Anne-Marie at the grotto. Carrel witnessed the euphoria of Anne-Marie and the reaction of the child when he began to see. There was no mistaken diagnosis this time. Yet Carrel was fixed in his materialist doubt.[91]

Alexis and Anne-Marie set up a home on remote St. Gildas Island off the coast of Brittany, where they enjoyed the beauty and peace of this refuge when they could. Anne-Marie spent most of her time with her son on St. Gildas, while Alexis pursued his career in New York.

COLLABORATION WITH LINDBERGH

Carrel's exploratory experiments on animals had sought to understand the relationships among organs. He understood a living body as an engineered system. He wanted to find a rationale for the particulars of the body. In unison the system of organs performed identifiable functions. He had no success in transferring an organ from one host to another, however. Carrel had been experimenting in this way for many years when he found a very talented collaborator, Charles Lindbergh, the hero aviator, who solved a major technical problem with Dr. Carrel.

In 1930 Lindbergh's sister-in-law, Elisabeth Morrow, was suffering from an inoperable heart condition. When Lindbergh asked her doctors why they could not surgically repair the damage to her heart, he was told that there was no way to repair a beating heart. The heart could not be stopped long enough to do the surgery.[92]

But Lindbergh knew the heart was a pump. So he asked why a mechanical pump could not perform the heart function long enough to do the surgery. Elisabeth's doctors did not know, but they knew who could answer

---

[91] Carrel, *The Voyage to Lourdes,* Real-View Books, p., 28
[92] Berg, A. Scott, *Lindbergh,* C. P. Putnam & Sons, New York, pp. 221-5

such a question: Dr. Carrel at Rockefeller. When Lindbergh met Carrel, he was enthralled by Carrel's very specific description of the problems involved and what had been done in that direction up to that time.

There were three issues that arose with heart pumps. First, blood coagulates readily on glass or metal surfaces. Second, blood vessels are delicate; they degenerate under the unmuffled force generated by pump valves. If these two problems could be solved, Carrel was sure that infection would become a problem. Asepsis would have to be achieved.

When Lindbergh saw the perfusion pumps[93] that Carrel had tried in earlier efforts to circulate nutrients through organs, he thought he could do better. He asked for permission to try. Carrel's assessment of Lindbergh was so positive that he gave him complete access to the laboratory's facilities. Their personal friendship and professional collaboration would last through the next decade, until World War II intervened.

Lindbergh came up with a succession of pump designs that worked first on simple tissue samples. When he had mastered the basics of tissue maintenance, he moved on to designing diffusion pumps that could bring nutrients to organs. Preventing infection, as expected, was a major sticking point. He worked diligently to master the underlying physiology and pathology as deeply as necessary to undertake the task. As often as Lindbergh came up with a new design, Carrel would schedule an appropriate experiment to test the prototype. A very deep mutual respect developed. Carrel marveled at Lindbergh's tenacity in study and ingenuity in design. Lindbergh relished the scientific rigor and optimism of Carrel.

Success came in the spring of 1935. Lindbergh had designed a perfusion pump, which Carrel used to successfully sustain a functioning thyroid gland from a cat for eighteen days *in vitro*. The final element in the successful design was a complex organ chamber that could be sterilized. The gland was attached within the organ chamber to a network of tubing and control

---

[93] A perfusion pump is a machine that will force a liquid from one point to another. A gasoline pump at a filling station is a coarse example of the process.

components that circulated the nutrient fluid at sufficiently uniform pressure to accommodate normal metabolism. Experiments on a whole series of organs, including hearts, followed. The Lindbergh pump concept, with the organ chamber, succeeded in maintaining organ function aseptically. A new era in the study of organ function and transplant surgery became possible.

Rejection of Descartes

Non-specialist Scientists

According to Carrel, the Cartesian philosophy of the Renaissance, which accepted only quantitative data as scientific, was a major mistake. He states: "We must radically differ from them and attribute to secondary qualities the same importance as primary qualities. We should reject the dualism (body and soul) of Descartes. The soul will no longer be distinct from the body. Mental and spiritual manifestations, as well as physiological processes, will be within our reach."[94]

Carrel's expanded type of science involves accessing the mental and transcendent in man through the body. Matter becomes the entry through which the mind and heart can be studied scientifically. He wants to avoid independent influences of the mind or the spirit. He will depend fully on the data of physical observation.[95]

His image of "the nonspecialist scientist" suggests the purity of heart of an unbiased, unselfish, altruistic yet penetrating mind, one that can infer from quantitative data the qualitative realities within. This is meant to be an inductive system, which parallels the always provisional rigor of physical science.

Carrel speculates that the training necessary to affect such altruism and wisdom would range between twenty-five years for a practitioner in a particular discipline and up to fifty years for the highest level of authority. He

---

[94] Carrel, *Man, the Unknown*, p. 279

[95] Ibid. pp. 280, 281

also estimates that it would take up to one hundred years to reach the level of maturity possible. Needless to say, he was not able to get funding for this idea. Carrel's venture into social sciences would continue in Nazi-occupied France under very difficult circumstances.

Obstacles

Carrel's unwavering scientism was unrealistic. If he had lived at the time of Pascal and Descartes in France, he would have seen substantial agreement among scientists on methods, despite Galileo's problems with the church. But he would have seen a changing philosophical outlook as Thomistic metaphysics was eclipsed. His nemesis, Descartes, did not foresee that his new philosophy would lead to idealism and weaken the grounding of science. Descartes drove the transcendent characteristics of being out of view, abandoning being to the mind of the experimenter. Today modern physics has changed this viewpoint. Today's physics involves the experimenter directly.

Carrel's tinkering with qualitative science proceeded on a limited scale. To infer qualitative characteristics by observing the quantifiable aspects of a human person is to some degree viable, but the repeatability and specificity of such "experiments" are problematic. Some nonspecialists may have a sensibility for qualitative inferences, but it would be close to guesswork. Within wide limits, very general conclusions might be possible, but the ontological gap between mind and matter is there, and it matters.

P.S. Doctor and Patient

In *Man the Unknown* Carrel revisits the dichotomy between the scientific view of man as an abstract human being and as a real person, the unique individual. Specific medical indicators apply to all human beings. But the physician only treats persons.

The task of balancing what is known of the patient as a human being (the object of interest to medical science) and as a person (a unique individual composed of body and soul) is complex.

As a physician Carrel had to know his patient's history, not only as a medical record but also how the story was presented by the patient. A good doctor recognizes that factors in one's personal life can affect bodily function and skew a physical examination. Objectifying the patient, while in one sense necessary, cannot be the only point of view. Carrel, as a scientific researcher, had to objectify what he studied. But as a generalist, his view of doing science on and about human beings was conditioned by human factors not subject to objectification.

The physician as an applied scientist has tools generated in response to human commonalities, the science. As a healer, however, he knows that disease intrudes uniquely in an individual and that the individual responds uniquely. So the question is: How does a doctor, possessing collective/general knowledge of man plus experience of and with this person, decide how to treat this unique patient? Inevitably, he must make prudential judgments. Carrel adds: "Medicine is not a discipline of the mind."

The omnipresent situation in each doctor-patient relationship recognizes that the individual patient *should never* be treated as a generic human being. The reductionist implications of medicine as a science must give precedence to the reality of the person, one created by God at conception for eternal life.

Chapter 6

Thinking as a Realist

Current Science and Philosophy

What follows regarding our understanding of relativity and quantum physics is complex. Do not be disturbed! While it may be difficult, be assured that it has a wonderful internal consistency. Each major step is associated with a physicist, most of whom you have met. In the spirit of Michael Faraday, stay with it. By staying with it, you will see its consistency and thereby, in the end, its beauty.

Modifications from Einstein's Relativity

Electromagnetism in the early twentieth century presented physicists with a set of facts that required them to drop the Cartesian philosophy of measurement. Special relativity requires the scientist to adapt his measurements of the speed of light, c, according to a prescribed process.

Measurements of space and time made in the reference frame of the experiment are made as usual and are called the proper measurements—proper length and proper time.

Measurements made in a reference frame moving at constant velocity relative to the proper reference frame must be adapted. In this reference frame, the measured time is dilated (enlarged) and the measured length is contracted (shortened). In addition, to ensure simultaneity, a multiplicity of synchronized clocks is required in the inertial frame.

These adaptations provide consistent experimental results. But the scientist participates in the adaptation of the measurements.

The observer outside the reference frame of the event is constrained to follow a procedure to adapt time intervals and space intervals in order to maintain a constant speed of light in his reference frame.

General relativity posits a further abstraction, a new geometry, a mass-induced curvature of space-time. The scientist now watches himself observing motion along curved lines in a new space-time geometry. The system has the potential for failure if the mass is large enough to create excessive curvature. In such a case, the solution of the general relativity equation produces indeterminacies called "singularities." The scientist's position as a subject in a space is conditioned by layers of theoretically imposed adaptations.

Quantum Physics and Its Modifications

Quantum physics primarily addresses interactions between and within subatomic particles, of which there are two types—those with mass (leptons), such as the electron, and massless particles (bosons), such as the photon. Quantum physics applies on the largest physical scale, the universe, and on the smallest scale. Astronomy provides significant data for particle physics and cosmology, the study of the universe, and in cosmogony, the history of the universe.

Quantum physics applies on a scale that approaches the lowest limits of size. In this limit, objects elude the grasp of familiar techniques, and at the end of the day they yield less than one would wish. Quantum mechanics can be a source of some distress when first encountered.

Recall that quantum mechanics arose from Planck's study of black body radiation. Planck "solved" the black body problem by assuming that each atom or molecule that was generating energy at frequency, f, was restricted to emit energy in little lumps, "quanta" called photons.

The energy emitted by an oscillator at frequency, f, had to be equal to an integer, n, times hf[96], *i.e.*, the energy emitted by a single oscillator had to equal nhf, where n = 1, 2, 3 . . ., and h is Planck's constant and f is the frequency.

So the total energy emitted from the body at one frequency was the sum of many oscillations of energy nhf. Such treatment of the data predicted the peak in the spectrum and also how the peak frequency varied with temperature. Why? How? No one, most especially Planck, had a clue.

Planck's quantum of energy idea sat for five years until Einstein, an obscure patent clerk, used Planck's description of light to explain the photoelectric effect. His straightforward and successful analysis depended directly on the assumption that light had a particle nature and that the photon, the light particle, had a quantum of energy, exactly as described by Planck. It took about twenty years for the implications of the Planck/Einstein nexus to be explored, broadened, and accepted.

Quantum theory received a third major boost in 1924 from a doctoral thesis written by a French graduate student, Louis de Broglie. His premise pushed Planck/Einstein quantum physics onto center stage. He proposed that just as we have found electromagnetic waves can have a particle nature, so also it is reasonable to assume that particles may be found to have wave properties.

De Broglie identified the wavelength, λ, of a particle with the momentum, p, of the particle. The Greek letter λ (lambda) traditionally designates a wavelength. For all waves the speed of the wave is equal to the product of the frequency and the wavelength, $v = \lambda f$.

---

[96] Planck's constant, h, is called the "quantum of action" because it is expressed in units of energy X time, erg-seconds, in the cgs metric system. In classical mechanics these units identify a physical property called action. The value of his 6.625 X 10^{-27} erg-seconds.

In classical mechanics, the momentum, p, of a particle of mass, m, moving with a velocity, v, is simply the product, mv, (p = mv). (At relativistic speeds there is a correction, but the correction is not important for our purposes.) By considering the units of measure of momentum, p, and wavelength, λ, and Planck's constant, h, de Broglie saw that the wavelength, λ, and the ratio, h/p, have the same dimensions.

Both sides of this equation, λ = h/p, have units of length. De Broglie's premise is in the same league with Planck's quantum postulate as far as boldness is concerned. These de Broglie waves came to be called matter waves.

The trio of scientists, Planck, Einstein and de Broglie established the fundamentals of quantum mechanics. These ideas wrought substantial changes in experimental understanding of measurement and the place of philosophy in measurement.

The Principle of Indeterminacy

Quantum mechanics' first major deviation from familiar experience is that it predicts an innate uncertainty in measurements made on the level of atomic sizes. This is familiarly called the uncertainty principle. Originally it had been called the principle of indeterminacy. I believe indeterminacy is a more appropriate word. In order to determine the position of an electron, we have to see it. To see it we have to illuminate it. The light energy needed on such a small scale disturbs the electron, making its position indeterminable and therefore uncertain. Here the act of measurement itself compromises the outcome.

In measuring on the scale of atoms, quantum theory provides us with a lesser alternative to the "certainty in principle" of classical physics. Quantum theory provides a set of all the possible outcomes of a measurement, and assigns a probability to each outcome. When a measurement is made, one of the allowed solutions will be found to be the case. And if you do the same experiment many times, you will find that the probability assignment is very good. The new theory works, but the physics is portrayed in a mathematical probability "space" and not in terms of a "subject-in-space-before-object" focus. All of the outcomes and their probabilities "exist" in a mathematically described set, called a Hilbert space.

Hilbert space is the "object" that is being observed by the subject, a fuzzy lens of sorts. Quantum theory requires that a measurement be made in order to reveal the state of the system. Measurement requires the subject to provide the stimulus to which the physical system can respond. The action of the putative nonintrusive subject is necessary to determine the state of the object. Without the action of the subject, the state of the object could not be revealed.

If the Cartesian model is to be maintained, the scientist must see his "nonintrusive" observation as subsisting in his mathematical knowledge of the Hilbert space within which his experiment has stimulated a particular result. He then objectively records which of the possible states from Hilbert space has emerged.[97]

Other Quantum Experiments
Interestingly, Wilhelm Roentgen, at the time, had been studying a new variety of electromagnetic wave—X-rays. X-rays were produced by slamming high-energy electrons into metals. The deceleration of the electrons as they are absorbed by the metal provides the energy emitted as X-radiation. This phenomenon was a large-scale inverse of the photoelectric effect, since very large numbers of mass particles, electrons, caused the emission of many photons, electromagnetic waves, from a metal.

Also, in the early 1920s, an American, A. H. Compton, conducted a series of experiments on the scattering of X-rays by electrons. He was able to show that the collision between an X-ray photon and an electron are

---

[97] The scholastic metaphysics which nurtured science from the 13th century through 17th century would recognize such complexity as an imperfection, or a hidden-ness of essence, as in complex structures like the atmosphere. Classical metaphysics can absorb the quantum dilemma.

dynamically the same as a collision between two billiard balls. Both types of collision obey the laws of conservation of energy and momentum. The X-ray energy losses appear as reduced photon frequency, supporting the Planck assertion that photon energy, E, depends upon the frequency, f. That is, $E = hf$.

Then, in 1927, Davisson and Germer at Bell Laboratories in New York reported experimental evidence that electrons exhibited wave properties, particularly that electrons could be diffracted by crystals. Diffraction is a wave phenomenon in which a beam of light is scattered in a predictable pattern when the beam is partially blocked by an edge or passes through an aperture or pattern of apertures. Showing that a crystal can cause electrons to diffract indicates that electrons have a wave nature.

CARREL IN RETROSPECT

The Unique Strengths of Youth

Looking at the progression of Carrel's life, it is clear that as a youngster he had very special talents, and he knew he did. For example, his special aptitude for surgery and his unusual manual dexterity are mentioned in regard to his early work as a medical student in the laboratory of Dr. J. L. Testut. Youthful anxiety is normal, but he may have foreseen objection to his expectation to do science alone, "his way."

The work he did that won him the Nobel Prize, fusing a torn artery, was done in 1912 when he was twenty-nine, two years after his graduation.

He was looking for a needle in a haystack—a position that would allow him to do what he wanted to do. But at no time did foreboding lead to relenting. He was "dead set" on being the unique scientist he imagined.

After realizing that his native France had no place for him, he migrated to Montreal in search of a suitable position. He struck out there too. He had an assistantship at the University of Chicago when Simon Flexner spotted him in 1906. Flexner had found a scientist uniquely suited to the mission of the Rockefeller Institute.

As a medical scientist, Carrel was what Lee Smolen[98] calls a "seer" rather than a business-as-usual "craftsman." The Institute had a commitment to new directions in medical research, which made it an ideal place for him. Winning the Nobel Prize ennobled his position. His manner was of no consequence during his premier days at the Institute. As long as Flexner was in charge, nothing was going to change. Flexner's choice had been brilliant. But it would have been fruitless for Carrel to look to his colleagues for approbation.

Ordinarily science is a team sport. But seers such as Carrel and Einstein do not join teams. He was, like many great scientists, a maverick. His grating manner called attention to his "outside-the-box" work ethic. His colleagues viewed his public visibility as self-promotion—unprofessional.

The Loneliness of Being Unique

In 1935 Carrel was denied an extension of his retirement age. He was being shown the door at Rockefeller, even as his mentoring was leading Charles Lindbergh to perfect the design of a perfusion pump, which ultimately would enable surgery on a human heart.

In that same year, Carrel published *Man the Unknown*,[99] a review of his scientific work and a very detailed description of what he saw for the future. The book sold almost a million copies worldwide. Yet we can readily understand why his vision of scientists who could infer states of mind and heart from physical examination did not draw support.

In terms of epistemology, Carrel had no regard for the boundaries Pascal saw among his three orders—the senses, the mind, and the heart. Much can be inferred about Carrel's behavior and thinking in terms of the three orders—particularly how the boundaries enable people to vary so widely in

---

[98] Smolen, Lee, *The Trouble with Physics,* Chapter 18, First Mariner Books, 2007
[99] Carrel, *Man the Unknown,* 1935

their decisions in life. For example, we have seen that in his sense of God, he struggled against knowledge of the third order, the domain of love. His inability, throughout his professional life, to resolve basic philosophical questions left him hanging on to threads—often providing explanations that only amplified the confusion.

For example, in a published report in 1902, Carrel was quoted as having stated, among other things, that a hopelessly sick Marie Bailly was radically cured. Being unhappy with the idea attributed to him, Carrel set the record straight by responding with a rambling letter to the editor:

> *For a long time, physicians have refused to study seriously the cases of miraculous cures. They have disregarded the fact that it is a huge scientific error to deny facts before examining them.*
>
> *In 1858 a shepherdess [Bernadette Soubirous] saw in a vision a person whom the Catholic religion identifies as the Virgin Mary. Many cases of {sudden} cures were observed among the sick {visiting in Lourdes}.*
>
> *We do not discuss beliefs, even though we risk thereby scandalizing both believers and unbelievers. We rather say that it matters little whether Bernadette was a hysteric or a fool, or even whether she really existed.*
>
> *What alone matters is to consider facts, inasmuch as they can be established scientifically, apart from all metaphysical interpretation.*
>
> *To many thinkers nothing can be produced except by the interaction of the forces of nature and within the range of facts long since observed. When there occurs a phenomenon, sufficiently*

recalcitrant to being enclosed within the rigid framework of official science, it is denied as dangerous because it breaks the customary formulas to which the human mind loves to be confined.

The so-called scientific minds deny such facts; others consider them supernatural. Later the supernatural side of the phenomenon may disappear together with our ignorance of the cause.

In the presence of new facts, we must limit ourselves to making exact observations and 'break the confines of philosophical and scientific systems as if one were to break the chains of intellectual slavery' (Charles Bernard).

One should be on guard against the fanaticism of sincere people and be ready to confront religious and anti-religious prejudices. Catholics must not consider scientific analysis as a sacrilege or an attack. Science has neither country, nor religion.[100]

Such a sorry parade of tentative nonsequiturs clarifies not at all.

Descartes's philosophy filled the void created by the misapplication of scholastic philosophy as interpreted by churchmen. The uncritical adoption of Cartesian ideas, by default, filled a vacuum, and for a while it managed to meet the requirements of classical physics without objection, but when Planck, Einstein, and de Broglie finished their work, the Cartesian model of scientific experiment was insufficient.

Qualitative aspects of reality before the Renaissance were described by Aristotelian metaphysics as well as religious faith. Recall again Pascal's

---

[100] Alexis Carrel, *The Voyage to Lourdes*, Harper Collins Publishers, reprinted with new introduction by Stanley L. Jaki, Real-View Books, 2007, pp. 19 and 20.

dialogue with Fr. Noel over the primacy of the senses in material matters. In a book titled *Prayer,* written in France in the early 1940s, Carrel suggests: *"We must listen to Pascal with as much fervour (sic) as we listen to Descartes."*[101] Pascal and Descartes are irreconcilable philosophically. They face to face, person to person, disagreed. Carrel, again, can't have it both ways.

A Case for Metaphysics

Philosophic Roots of Science

Science emerged in the late Middle Ages in Europe. Fourteenth-century Europe was a culture whose people had believed in a benevolent, personal God who created all there is. Long-standing conviction on the objective existence of things supported three presuppositions necessary for doing science:

1. There is order in nature;
2. Human beings are able to detect that order; and,
3. Detecting the order in nature is a good thing to do.

These are simple philosophical suppositions—one metaphysical, one epistemological, and one ethical. They grounded medieval science in realism. The identity—with a loving, personal Creator—centered European culture in realism. Indeed, realism is announced by the tangible character of the words in the Creed, which refer to creation, conception, birth, suffering, death, resurrection, ascension, judgment, etc. When science started in Europe, the pioneers, before and after Copernicus, were unfettered in mind regarding the reality of the external world.

Whether Galileo intentionally demeaned his clerical critics or not, scientists of his day knew that his astronomy was based on compelling physical

---

[101] Carrel, Alexis, *Prayer,* Trans. Dulcie de Ste. Croix Wright, Morehouse-Gorham, N. Y., 1948, p.54.

sense. The astronomic observations of their time strongly suggested that the centrality of the sun was a more fruitful working hypothesis. The harsh treatment of Galileo did make scientists chary of open conflict with the Church. Descartes prudently delayed publishing his *Treatise on the World* in 1633 to avoid censure. But Catholic scientists were not paralyzed.

Blaise Pascal was ten years old when Galileo was censured. Fourteen years later, in 1647, when Pascal published his work on the vacuum, he too was challenging Aristotelian physics. He defended his work against the philosophical invocations of Father Noel. In a very cordial exchange, Pascal upheld the exclusive use of experiment in the determination of what is physically true. Philosophical arguments, which Pascal recognized in matters of the mind, had to defer to sense knowledge in matters physical. Science did not need Descartes's *Cogito*, which at that point was in the air for ten years. Pascal rejected the *Cogito* statement as bad philosophy from the outset.

The Ontological Structure of Being

The Aristotelian metaphysical tradition imputes our knowing to the innate knowability of objects, which knowability is proffered to the mind, which forms an idea of the object. The man knows the object by means of his idea of it. The idea depends upon a projection from the being itself to the mind. Sometime in our forgotten childhood, each one of us experienced, for the first time, the intrusive linkage between an object and its name. There is one scenario that indelibly declares this linkage.

Kenneth Schmitz describes the process:

> *There is at least one exception to the forgetfulness of the discovery of sign and concept, and that is the remarkable experience of Helen Keller, corroborated by her teacher. In her autobiography Helen recalls how, when she was a child, deaf and blind, her teacher pressed the sign for water into her hand*

> *as cool water splashed over her fingers. The connection between sign (word) and the thing (water) was instantaneous, as was her response. She fell upon the ground one hand clutching the teachers, the other clutching objects, one after another, to demand from the teacher the sign (word) that named each. From that moment of the discovery of literacy, Helen developed into the fully intelligent and educated person she became. With the discovery of sign and concept, the dimension of human thought opened up to her.*[102]

The "sign for water," the word, was linked with the tactile sense of water itself. Such is the means by which our mind is grasped by an object.

Science is built on the presumption of constancy in what is physically measurable in a being, or a process. Much can be inferred from this presumed constancy. It suggests an implicit trust in a level of order in the universe. What does this imply about the world?

The Cartesian view distorts by locating knowledge of the objective world within the observer rather than the alternative of recognizing that physical beings, as beings, are innately knowable. Each being proclaims its uniqueness, and we are capable, in many cases, of grasping it. It cannot be that the truth of a being does not exist if we don't know about it. It is ridiculous to insist that a being only exists when we come to know it.

Transcendent Characteristics of Being

Let's look at a metaphysical approach that will both ground science and relieve the problems Descartes has with relativity and quantum mechanics. The metaphysics used here is adapted from the same work by Schmitz.[103] He presents this metaphysics in the context of human anthropology. His

---

[102] Schmitz, Kenneth L., *Person and Psyche,* Institute for Psychological Sciences Press, Arlington, Va., 2009, p. 44

[103] Ibid., Ch. 1, pp. 1-16.

elucidation of his metaphysics is preparatory to a discussion of psychotherapy, but it carries over very well to physical science.

The major insight of this metaphysics is that there are six necessary transcendent characteristics present in every being as an existent. *Necessary* means that every being must have them. These characteristics are found in every being analogously, i.e., in a manner appropriate to the being it is. Five were recognized by Aristotle in ancient times. Aquinas is responsible for the sixth.

These six characteristics are:

(1) Every being has what is called a *form*. Another word for form is *essence*. The form of a being is what makes the being what it is. This characteristic, like all the others, has to be seen as arising from a necessary order, or presence, that inheres in the act of existence. Every thing has to have a mixture of specificity and variance. Form focuses on elemental specificity—what is it?

(2) Every being is a *unity*. When we perceive a being, it is always a particular being—a unit. It may be found in some complicated context, like yeast, a collection of single-celled fungi, in a cake. But the yeast is in there as yeast, necessary in the recipe. Each fungus constituting the yeast gives up its order in baking, but to the extent that each fungus exists within yeast it also is one.

(3) Every being is active in its relationship with other beings. Beings proclaim their *otherness*, its alien status, to other beings. We can say each being is in active possession of its existence.

(4) Every being is knowable, *intelligible*. The intelligibility of a being resides in the being itself, not in the knower. The human thinker in knowing a being has accessed the ontological truth, or meaning,

of the being. In the process the knower has realized, i.e., made real the intelligibility of the being. Yet intelligibility is a property of the being whether it is known or not. When the being is known, its intelligibility is realized.

(5) Every being has value that resides in its being as a *good*. In knowing the truth of being, we are able to access the good that resides in every being. These *goods* can be used to enhance our human existence.

(6) Finally, Aquinas added the property of *luminosity* to Aristotle's original five transcendental properties. Our human realization of the luminosity in being is manifested as *beauty*.

Together these characteristics constitute what Schmitz calls the *texture of being*. Every thing that exists has each of these characteristics. To be is to have all six. Transcendence refers to the necessary presence of the characteristics in every being.

The analogous nature of each transcendent characteristic consists in its being present in a manner appropriate to the being. Each of the six characteristics inheres in a particular being in ways appropriate to that being—i.e., it inheres in a manner consistent with the form of the being, what it is.

This texture of being gives us a sturdy and fertile platform from which to analyze being in general and particularly man's relation with other beings. A primary, immediate implication from form, unity, and otherness is that the inclinations with which the human being, as well as every other being, is endowed must be ordered to the *good* of the being. These endowments, the nature of the being, must support the fulfillment of the being. Otherwise, there is an untenable contradiction in being, a lack of ontological goodness. Therefore, every being can be said to be good, analogously, in a way appropriate to the being.

Metaphysics expands our vista of being. Right away it recognizes a necessary coherence between our fundamental endowments and the fulfillment of our humanity. Our endowments must be directed toward our good. Similarly, every other being is directed toward the good of that being, according to the kind of being it is.

We recognize that the good of one being may not be good for other beings. No one likes to be in the neighborhood of a skunk, but the odor of a skunk works for the good of the skunk—not anything else. It's a "skunky" good, appropriate to the well-being of the skunk.

Schmitz, in order to emphasize the tangible element in being, uses the term *actuality of presence* to describe the manifestation of being and to dispel the vague universality implicit in the term *being*. He insists that the polar opposite of being is nothingness, and suggests the term *omnipresence* to guard *being*'s universality while ensuring its actuality.

However, we encounter being often hidden in a mélange of presences. Being is often manifested with imperfections, distortions, and privations. Being is rarely framed in a pristine context. To recognize the mix of negativity that tarnishes the being we often experience, Schmitz introduces the term *reality* to describe being so tarnished. Reality is being fraught with absences. The efficacy of the term *reality* is inferred by Schmitz's context. He is presenting this metaphysics anthropologically, preparing to discuss its relevance in psychotherapy, which is concerned with privations of being.

A concrete scientific case of a multiplicity of presences is meteorology. The mélange of factors that influence the atmosphere at any given time is problematic. Reality is a weather map with isotherms and isobars, currents of air and water, dew points, highs and lows, etc., that attempt to give coherence to the atmosphere as a being. The weather reports are, as you know, informed guesses.

Yet no matter how limited, each being has transcendental qualities, an ontological structure, despite defects or privations. In its essence, each being is a unique enclave of order, a prescription that enters into the absolutely radical makeup of every reality. It is precisely this metaphysical prescription that grounds sense knowledge in reality. This is a very important consideration for the epistemological integrity of science.

Being, therefore, has a foundational framework that is coherent and affirms itself. Every being, in itself, fundamentally possesses its own being, with all the transcendent characteristics of same. An original order is established within every being, which enables that being to participate with other beings by virtue of their individual transcendental characteristics.

This fundamental possession of being can be seen in the behavior of an electron, a fundamental subatomic particle. An electron has three measured characteristics—mass, charge, and spin. Its mass is small, which shows that the electron is weakly influenced by gravity. The electron has a negative electrical charge, large enough to enable it to play a primary role in chemistry, the dynamics of atoms and molecules.

In addition, the electron responds to a magnetic force through another measured property we call its *spin*. The term *spin* should not be literally interpreted. Just think of spin as a little arrow that can point in one of two ways. The electron spin can line up in the direction of a magnetic force or down.

Physicists have found that the electron spin has an odd behavior called entanglement. If a pair of electrons is coupled together at some point, their spins must be aligned either parallel or anti-parallel, i.e., in the same or in opposite directions.

Now, we find experimentally that when two entangled electrons separate physically, they maintain their spin orientation as they move away from each other. No matter how far they separate, they maintain their orientation. This phenomenon is baffling, especially over large distances. How can two widely separated electrons remain entangled? Experiments show that they do!

You would have to say an electron has a natural inclination, a fundamental possession of its own being. Note that the electron's otherness, its principle of acting, is letting scientists know that it is entangled. This is a surprising and anomalous reality, which has still to be adequately absorbed by the physicists. Note, also, that scientists, beings capable of knowing, are interested in electron behavior and that the electron reveals itself to them. Thus, they know the electron to that extent, and in the future this knowledge may be used for some good purpose. The metaphysical system of Aristotle is fine with this.

In summary, each being, in ways appropriate to itself, is ontologically inclined to realize its given nature. These orders are operative in every being, structured from basic universal orders in simpler beings to increasingly more specialized expressions in higher levels of being. These ontological orders, the transcendental qualities of being, secure the most fundamental possession of being.[104]

Every being has the innate tendency to express, i.e., bring to reality, its given endowments. In a living being we would call it an innate drive, which compels assertion of its own being. Most simply, every being is constituted to realize its nature, i.e., to act and interact according to its nature, what it is—to radically possess and proclaim what it is. Granted the range of natures to be realized is enormous, and for this reason, some choose to dismiss this insight as bland and unfruitful.

Geologists analyze rocks. Your backyard becomes very interesting if you know a little geology. Every rock has a unique composition, texture, size, shape, hardness, location, etc. All its properties determine both that it is a rock (essence) and that it is this rock (unity), not any other rock, but it can interact with other beings (activity). These determinants enable this rock to communicate its being by interacting within its environment as appropriate for a rock.

How does a rock be a rock? Consider a rock in a mountain spring stream. There is a giving and a receiving between the rock and the water. The rock

---

[104]Draft of Human Nature and Human Culture, in Fall 2008 was principal source. The finished Human Nature and Human Culture was later published in Journal of Law, Philosophy and Culture in 2009, Vol. III, No.1 (2009) pp. 87-

is gently abraded by the water, giving it a smooth surface, which landscape gardeners like for rock gardens. In return, the rock gives the water a unique mineral content. As you know, the mountain spring water can be bottled and sold in the supermarket. The landscaper gets rocks and refreshment from the deal. Both the rock and the water contribute to each other in a manner appropriate to their being. In addition, they provide goodies to their human neighbors and are part of our endowment.

FREEDOM AND CULTURE

The Person and Practical Reason
Thomas Aquinas, when describing the purposes for which man addresses the variety of being before him, recognizes a preference for pragmatic utility in how man uses the good available in being. Being is first used to provide for fundamental human necessities—food, shelter, clothing, families, community, etc. To wit:

> *Now a certain order is to be found in those things that are apprehended by men. For that which first falls under apprehension is being, the understanding of which is included in all things whatsoever a man apprehends.... Now as* being *is the first thing that falls under the apprehension absolutely, so* good *is the first thing that falls under the apprehension of the practical reason, which is directed to action (since every agent acts for an end, which has the nature of a good).*[105]

Though the syntax may be awkward, the gist is that man does not in the first instance think philosophically. Rather, he uses the beings presented in his world first to provide for basic physical needs: food, shelter, clothing,

---
[105] Aquinas *Summa Theologica*, I-II, 94,2

and implements for the comfort and defense of the person and his community. In the process he acquires an understanding of being that is specific to him. But he also develops a common level of understanding with others in his community.

In his encounter with being, he from the beginning has an inherent ability to transcend the limits of his bodily senses. In his self-conscious perception of being, there is an awareness of the otherness of being, be it animal, vegetable, or mineral. This is an effective awareness of relationship within being, an inclination to infer, not formally but effectively, the qualities of form, unity, and externality of other being. The human intellect works in the total context of human personhood—physical, mental, and spiritual.

Schmitz[106] identifies three components in human personhood. First is the body—the totality of biophysical and biochemical systems materially present, which are organized to implement all the physical capacities of the person.

The second component, sensory life, concerns a human phenomenon shared in a lesser degree with higher animals, i.e., sensory perception and its relation to emotional response. In the scientific realm, the primary interest is in the physics/chemistry of sense experience and the ontological status of emotional response. Imagine a person hearing a few bars of music. We have a good understanding of how the ear is conformed to distinguish sound waves of different frequencies. The ear's transmission to the brain generally elicits a response in the person.

The response, however, arises not from determining mechanisms but from the consciousness of the person. Consciousness hears music not as a sequence of waves but as a melodic whole, which is more than its physical characteristics. The physical state of the hearer's brain at the time depends upon many things, but the sense perception, hearing, activates the hearer as an agent, one who acts, not as a physical network of neurons. While her

---

[106] Schmitz, *Person and Psyche, p.38 ff.*

brain is always in some physical state, quite simply the physics of sound is not what she hears. The person hears music. She, as the person she is, will determine her emotional response.

To view the brain, influenced by the physical train of waves, as determining her response is a starkly Cartesian assumption. The hearer is not a Cartesian object but has the ontological character of an agent. She is not a stolid recipient of sonic energy. There is a meaning to the music beyond the physical. Her agency is a primary function of her personhood. Personhood is opaque to the physical sciences but opens itself to psychological and spiritual sensibilities.

We have alluded to the third component of personhood in the preceding discussion of the practical reason. At some point in life, relatively early, man realizes his ability to reason. He acquires a capacity unique to man, the ability to grasp the essences of beings and by means of language to communicate ideas. The emergence of language enables the mature person to grasp essences and associate words with things. The word can express a class of beings with a commonality, which the intellect can conceive as form.

When, through the intellect so constituted, the workings of the material world are understood, man has transcended dependence on senses. His intellect can probe the fullness of what being has presented to him. Consciousness is freed to enter into examination and exploration of the truth of things—how the world works. He can imagine and remember and conceive. He can plan and plot, design and execute. Mastering the truth of being expands the reach of his mind and can create new solutions as well as improve on existing ways of doing things. These capacities provide him with choices. The human apprehension of the good in being is the first step that reveals the spiritual elements in man—first knowing through intellect and then choosing through the will; first knowledge and then personal responsibility

One can see this progress in toddlers, who every day probe their world for things that interest them. They interact sensually with being from day one. Toys are beings that attract interest and concentration and then stimulate laughter. Delight is soon enough filed away, and each toy is thrown away when it is familiar. New beings are found and fondled. Soon things have names and words are used. The body's senses are augmented by knowing, remembering, imagining. The intellect begins to develop in terms of a language. The youngster is on a faster track.

All these events happen in a context that preexists. Descartes is nowhere in sight. The child is; therefore, he thinks,—exercising innate capacities through beings around her. This process leads to maturity and, ultimately, responsibility—all through the good, the beautiful, and the true found in being.

The person, when mature, is an actor who plays on a stage built and bequeathed by his forebears. When maturing, culture provides a context within which he will exercise his practical reason in attending to his fundamental needs. His culture has been generated by human ontological inclinations, the transcendental characteristics of being, which inhere in man and analogously every other being. The good, the beautiful, and the true are the generators of human knowledge and freedom. They incline human beings to create and contribute to culture.

The Culture as Context

The main inherited features of the cultural scene are (1) a language, which includes literature, oral traditions, and a history; (2) a social structure whose primary aim is to promote the traditions of the society and provide customs that assure peaceful order in common human pursuits; (3) an economic system that enables satisfaction of basic human needs and provides avenues for personal fulfillment and ways to contribute to the progress of the culture; and (4) a system of societal discipline, i.e., law, that channels human conduct within the bounds of justice.

The culture influences human consciousness in time, creating a striving to freely realize chosen objectives. In any creative period, the human being is aware of time, for he can use the fruits of time past and he foresees future time. As he traverses each creative period, he is directed by a conscious pursuit of an objective—a course of study, economic independence, courtship, adapting to good fortune or bad, contributing to the community, etc. In each such period man uses his freedom to access the good in being for the goals chosen. These individual pursuits constitute a process of willful and effective choice, for well or ill.

The person, in freely acting to access the good in being, first probes to find or extend his knowledge, the truth of being, then uniquely accesses the good therein and produces the desired solution to the perceived need. He freely chooses to use that good to satisfy the need. His culture is the context that frames his work.

It is important to note that the word *need* has variants, e.g., *want* or *desire*. Obviously, *want* and *desire* have different connotations than *need*. A man's revelation of what he needs can be an insight into his character. No man chooses evil as evil. Rather man addresses needs that, in his context, are good—maybe selfish or even criminal but, in his context, good. In these personal choices man reveals his own standard of the "good." The totality of a person's acts reveals his moral perspective.

The culture itself acquires a character from (1) the integrated perspectives of "good" in its people and (2) the culture's acceptance of the consequences of these perspectives. In this arena, a moral judgment of what is "good" is implicitly chosen and becomes a measure of individual human and cultural character. Human freedom derives from the transcendental properties of being, the good and the true, and also, on a not so pragmatic level, the beautiful. The freedom to choose reveals the "good" of the human actor and establishes a basis for assessing moral character.

Thomistic Realism
This classical adaptation of Aristotle's metaphysics anchors the physical world in realism. The knowability of a being is an inherent property—one

that exists in the being independent of whether it is known or not. There is no imposition of any kind on a being when it becomes known. Rather, an innate potential of the being, the transcendent characteristic of knowability, has become realized. Being known is an intransitive relation, not intruding on the object known. The being is knowable, whether it is known or not.

Knowledge endows the knowing human being with the truth of the known being. The knowledge occasions a new freedom to interact with the known being in pursuit of the good in that being. The freedom of the person resides in his choice of ends for which he will make use of the good in the known being. The realized goodness can solve a practical problem or achieve a benefit previously unavailable. Sometime, somewhere, someone had the intuition to bake the first apple pie. We are all the beneficiaries. The free use of this power to access the good of being is the primary way by which we influence culture.

The tendency in modern man to look inwardly, to focus on oneself, is very strong. Attributing the reality of the physical world, as determined by experiment, into the subjective idea(s) of scientist(s) is a philosophical overreach. The intelligibility of the known being is an invariant ontological proclamation of its reality. The being exists as it did before being known. This non-Cartesian status of being, as independently existing, grounds physics perfectly.

The inviolable inherence of knowability within a known object of a given genre will reveal the same experimental result when repeated, within the limits determined by the experimental conditions. Scientists assume and depend upon this constancy. The unchanging knowability of the object known is the metaphysical grounding of science.

The experimental data developed by physicists is most often, but not always, quantitative. In chemistry, processes are observed, understood, and controlled, again with quantitative consistency. Also, in biology, characteristic processes and structures are repeatedly observed in given species. No matter which type of data is developed, the expectation and the experience is that these results are repeatable or variable within understood limits.

Michael Faraday produced original and unusual descriptions of electromagnetic effects—fields and field lines, flux, and flux density, which, although remarkably intuitive, are totally effective in describing electromagnetism. The primary characteristic of all established modes of scientific understanding is that they are repeatable. This characteristic, a consequence of the inherent knowability of being, grounds science in reality.

In quantum theory, the knowing subject and the knowable object indeed interact directly physically. This interaction compromises the presumed Cartesian isolation of the subject. A realist epistemology does not remove quantum indeterminacy or the presence of an observer in relativity theory. It just does not forbid them. The problem does not arise. The metaphysical understanding of the role of the being studied, the object, and the knowing mind, the subject, allows an interactive relationship, without compromising their individual essences or the transcendent characteristics of their being. There are further insights that arise on the quantum scale, but the problems of current physics are down that road.

Afterword

We hope we have conveyed a sense of the history, breadth, value, and challenge of science and the central role of experiment. Theory is strengthened by falsification—the scientist's efforts to test a theory's truth. Also, a primary place must be given to metaphysics—literally, what is beyond physics. Recall the three simple propositions:

1. There is being before us. (Metaphysics)
2. We are able to know it. (Epistemology)
3. It is a good thing to do. (Ethics)

We have given a sense of the state of physics throughout time while including relevant philosophy. We have included the thought of individual scholars and, in some cases, their radical differences: Blaise Pascal and René Descartes, Alexis Carrel and Charles Lindbergh, Isaac Newton and Wilhelm von Leibnitz, Nicholas Copernicus and Galileo Galilei, Michael Faraday and James Clerk Maxwell, Max Planck and Albert Einstein. The metaphysics of Kenneth L Schmitz and its application to psychology in *Person and Psyche* is a unique recent development.

The title, *Finding the Light: Science and its Vision*, is a reference to the philosophy of Blaise Pascal as presented in the beginning of chapter 5. Pascal's philosophy is an epistemology, a unique philosophy of knowledge. According to his system, there are three orders of knowledge. First is sense knowledge, acquired by the body. The senses present something to the mind, the second order of knowledge. This is the level at which science is done.

Clearly, the mind of most of us does not engage in science. Most often the mind is directed to practical applications unique to the person. This is the level of most human activity—how bridges are built and cards are played, etc. The highest level of knowledge is of the heart, what is known to the person by virtue of wisdom. Pascal's faith was at this level due mainly to his Memorial Experience on Monday, November 23, 1654 (see *Describing the Orders* in chapter 5).

The practice of science is never absolutely complete. Every scientific theory is subject to falsification. Indeed, we have seen that some physicists, among them Lee Smolen, believe that in our day we have made a major error somewhere.

We have tried to stay within the history and the progression of knowledge that has brought us to this day. We would be remiss, however, if we did not acknowledge the current militant atheistic attack by several scientists seeking to promote atheism through their rather selective interpretation of science. Some recent authors are:

Richard Dawkins, *The God Delusion*.
Victor Stenger, *God, the Failed Hypothesis: How Science Shows That God Does Not Exist*.
Taner Edis, *The Ghost in the Universe*.
Christopher Hitchens, *God Is Not Great*.

In addition, Emile Zuckerkandl, a molecular biologist; Steven Weinberg, a primary theoretical physicist; Peter Atkins, an Oxford theorist; and psychologist Steven Pinker are among those who have weighed in on opposition to religion and the idea of God. Richard Dawkins's book has been very successful, leading many to identify with his unbridled call for an end to theism. While the materialism of Western culture resonates with the idea of freedom from religion, there is some good news—*Deus est*.

Afterword

Perhaps the most striking blow to atheism is the publication of *There Is a God*[107] by philosopher Antony Flew, for many years the most ardent advocate of atheism. Consider the following from the beginning of chapter 10 in Flew's *There Is a God*:

> Science qua science cannot furnish an argument for God's existence. But the three items of evidence we have considered in this volume—laws of nature, life, with its teleological organization, and the existence of the universe—can only be explained in the light of an Intelligence that explains both its own existence and that of the world. Such a discovery of the Divine does not come through experiments and equations, but through an understanding of the structures they unveil and map.

In appendix A of *There Is a God*, Roy Abraham Varghese provides a response to several new atheists: Daniel Dennet, a philosopher of science; Lewis Wolpert, a biologist; Sam Harris, philosopher and neuroscientist. In *The New Atheism: A Critical Appraisal of Dawkins, Dennet, Wolpert, Harris and Stenger*, he identifies five phenomena ignored by the new atheists:

1. The rationality of the physical world.
2. Autonomous life.
3. Consciousness.
4. Conceptual thought through languages.
5. The self as a center of consciousness.

---

[107] Anthony Flew with Roy Abraham Varghese, *There Is a God: How the World's Most Notorious Atheist Changed His Mind*, Harper One, 2007.

Another comprehensive rebuttal to the new atheism is David Berlinski's *The Devil's Delusion*.[108] In addition to providing insight into the materialism of the new atheism, Berlinski is a source on the current status of Darwinism, historically a biological issue. Berlinski, a secular Jew, does not write from a theistic sensibility but from a career "in studying mathematics and writing about the sciences."[109]

Although there are several sections in *The Devil's Delusion* in which Berlinski targets Darwinism, one stands out.[110] Entitled "Dull, Dutiful, So Very Darwin," he ends by quoting an unidentified Nobel laureate in biology: "Darwin? That's just the party line."

Berlinski (p. 185) highlights a seemingly breathless tirade by Emile Zuckerkandl castigating theists for promoting "Intelligent Design," a concept introduced scientifically by Michael Behe, a biologist at Lehigh University, in his book *Darwin's Black Box*[111] in 1996. Behe describes the mechanism involved in the functioning of a particular element in many bacteria, the bacterium flagellum (pp. 70–73). Figure 3-3 on page 71 is a visual display of the complexity of the system. In closing (p. 73), Behe comments: "As the number of systems that are <u>resistant to gradualist explanation</u> mounts, the need for a new kind of explanation grows more apparent. Cilia and flagella are far from the only problems for Darwinism."

Thankfully there are many accomplished scientists who are believers. Francis S. Collins, who headed the Human Genome Project, is a Christian. He is also a Darwinist, who faithfully represents the culture of biology in America. He has written a book, as a biologist,[112] in which he presents his "evidence for belief," which is very moving and commendable. However, in his writing, he seems not to recognize the very strong effect philosophy has

---

[108]David Berlinski, *The Devil's Delusion*, Crown Forum, 2008

[109]Ibid., p. xi

[110]Ibid. pp.185 192

[111]Behe, Michael J., *Darwin's Black Box*, Touchstone, 1996

[112]Collins, Francis S., *The Language of God*, Free Press, 2006.

had on science, particularly René Descartes's philosophy. See physician and biochemist Leon R. Kass's *Life, Liberty and the Defense of Dignity*.[113] Kass devotes the final chapter of his book to the catastrophic effect of Cartesian objectification of being in biology.[114]

In his *The Language of God*, Collins responds to Michael Behe's science under the category "Intelligent Design"—ID. He begins his response to Behe by recording the history of various people he identifies with the "ID movement.[115]" After this ad hominem précis, Dr. Collins suggests: "If the logic truly had merit on scientific grounds, one would expect that the rank and file of working biologists would also show interest in pursuing these ideas, especially since a significant number of biologists are also believers."[116] Can any fair-minded realist imagine rank-and-file biologists revealing an open mind on Darwin? Not if they want tenure! Lack of support from rank-and-file biologists is to be expected.

W. Norris Clarke, in his *The One and the Many*,[117] chapter 10, "The Metaphysics of Evolution," comments on the difficulty of and the unacceptably high odds against, an explanation of transition points in emerging life without intelligent design. Certainly, Clarke's reasoning is philosophical, free of scientific influences.

Behe's most recent book[118] was published in 2007. Let us simply look at the intensity and breadth of media response to Behe. The *New York Times*, on Sunday, July 1, 2007 (a Sunday edition has the greatest circulation) printed, and Richard Dawkins himself wrote, a review ("Inferior Design") that skewered Behe. Dawkins condescendingly and joyously celebrated the

---

[113] Kass, Leon R. Life, *Liberty and the Defense of Dignity*, Encounter Books, San Francisco, 2002, See Chapter 10 for a in depth discussion of the philosophy of science.

[114] Ibid. see Chapter Ten, *The Permanent Limitations of Biology*, pp.277 to 297

[115] Collins, pp. 181-186

[116] Ibid. p. 18

[117] Clark, *The One and the Many*, Ch. 2, p. 5 and p. 255

[118] Behe, Michael J, *The Edge of Evolution*, Free Press, 2007

isolation of Behe from his colleagues at Lehigh and from the other biologists. Dawkins and Co. has no place for a maverick.

Suggested Reading

Stanley L. Jaki's *The Savior of Science* has a meticulous treatment of the distractions used by atheistic materialists to evoke responses to empirical declarations.[119] What are meaningful, in argument, are the inferences that can be drawn from observation metaphysically, beyond physics, such as "causality, purpose and freedom of the will." Metaphysics is not about physical observations "but about inferences from them to an unobservable realm."[120] Attempts to forego what the mind can infer lead to tautological slogans such as "the survival of the fittest."

Jaki indicates a second trait of Darwin's materialism—its promulgators behave as if their knowledge were "catechismal." A catechism is "a summary of religious doctrine usually in the form of questions and answers."[121] Darwinists have a catechism. Stephen Jay Gould, an eminent paleontologist and gifted writer, on July 19, 1987, wrote an article in the *New York Times Magazine*, "The Verdict on Creationism." He comments (p. 34) on the status of his science: "Our continuing struggle to understand how evolution happens (the theory of evolution) does not cast our documentation of its occurrence into doubt." You see, Darwinism is documented; it is in their catechism. Only if the theory is well-established fact would the statement on the troubles with the theory make sense. But the catechism says evolution is a well-established fact, so the "struggle" is minor and will be worked out.

Jaki's third contention is that Darwinism has a choke hold on biological communication media—leading departments of biological sciences, leading publishers and editors of monthly, weekly, and daily publications. We have

---

[119] Ibid. See Chapter 6 *The Creator in the Dock*, particularly pages 204–206
[120] Ibid. p. 204
[121] Webster's Ninth New Collegiate Dictionary.

mentioned this problem previously in the context of biology departments. The prejudice is much wider.

Stephen Barr's *Modern Physics and Ancient Faith*[122] is a very detailed historic account of the interplay of physics and theology up to the present time. Barr is a professor of theoretical physics at the Bartol Research Institute at the University of Delaware. He presents an extensive discussion under five headings: The Conflict between Religion and Materialism; In the Beginning; Is the Universe Designed?; Man's Place in the Cosmos; and, What is Man?. Barr's work is very thorough and recommended for depth and breadth of scholarship.

C. S. Lewis in *The Abolition of Man*,[123] describes this same philosophical disease in the context of an elementary school textbook designed to teach English usage. Taking a Cartesian view of literary matters destroys the capacity for nuance, allegory, and other elements of literary criticism, never mind poetry. Lewis develops the frightening implications of this philosophy for man and his culture.

On Darwinism, George Sim Johnston's *Did Darwin Get It Right?*[124] is an overview of Catholicism and Darwinism. This book is aimed at the general reader but does contain items of scholarly interest.

I owe a special thanks to the Library of Pace University in Pleasantville, New York, the Georgetown University Library in Washington D.C., and to the Library at Ave Maria University in Naples, Florida. In addition, I spent many hours in the Public Libraries in Rye, Harrison, and Port Chester, New York.

I want to thank my colleagues in the monthly meetings of the Pace University Discussion Group who have presented me with a broadened view of philosophy, mathematics, and physics: the late W. Norris Clarke, SJ, Ernie Sherman, Yves Martin, Peter Knopf, Mohsen Shiri, and Steven Rosen.

---

[122]Barr, Stephen M., *Modem Physics and Ancient Faith*, U. of Notre Dame, 2003
[123]C. S, Lewis, *The Abolition of Man*, Harper One, 1944
[124]Johnston, George Sim, *Did Darwin Get It Right?* Our Sunday Visitor, 1998

CPSIA information can be obtained
at www.ICGtesting.com
Printed in the USA
BVHW041925200223
658865BV00010B/119/J